Communications
in Computer and Information Science **1103**

Commenced Publication in 2007
Founding and Former Series Editors:
Phoebe Chen, Alfredo Cuzzocrea, Xiaoyong Du, Orhun Kara, Ting Liu,
Krishna M. Sivalingam, Dominik Ślęzak, Takashi Washio, Xiaokang Yang,
and Junsong Yuan

More information about this series at http://www.springer.com/series/7899

Jian Chen · Van Nam Huynh ·
Gia-Nhu Nguyen · Xijin Tang (Eds.)

Knowledge and Systems Sciences

20th International Symposium, KSS 2019
Da Nang, Vietnam, November 29 – December 1, 2019
Proceedings

Springer

Editors
Jian Chen
Tsinghua University
Beijing, China

Gia-Nhu Nguyen
Duy Tan University
Da Nang, Vietnam

Van Nam Huynh
Japan Advanced Institute
of Science and Technology
Nomi, Japan

Xijin Tang
CAS Academy of Mathematics
and Systems Sciences
Beijing, China

ISSN 1865-0929   ISSN 1865-0937 (electronic)
Communications in Computer and Information Science
ISBN 978-981-15-1208-7   ISBN 978-981-15-1209-4   (eBook)
https://doi.org/10.1007/978-981-15-1209-4

This Springer imprint is published by the registered company Springer Nature Singapore Pte Ltd.
The registered company address is: 152 Beach Road, #21-01/04 Gateway East, Singapore 189721, Singapore

Preface

The annual International Symposium on Knowledge and Systems Sciences aims to promote the exchange and interaction of knowledge across disciplines and borders to explore the new territories and new frontiers. With over 19-year continuous endeavors, attempts to strictly define knowledge science may be still ambitious, but a very tolerant, broad-based, and open-minded approach to the discipline can be taken. Knowledge science and systems science can complement and benefit each other methodologically.

The First International Symposium on Knowledge and Systems Sciences (KSS 2000) was initiated and organized by Japan Advanced Institute of Science and Technology (JAIST) in September of 2000. Since then KSS 2001 (Dalian), KSS 2002 (Shanghai), KSS 2003 (Guangzhou), KSS 2004 (JAIST), KSS 2005 (Vienna), KSS 2006 (Beijing), KSS 2007 (JAIST), KSS 2008 (Guangzhou), KSS 2009 (Hong Kong), KSS 2010 (Xi'an), KSS 2011 (Hull), KSS 2012 (JAIST), KSS 2013 (Ningbo), KSS 2014 (Sapporo), KSS 2015 (Xi'an), KSS 2016 (Kobe), KSS 2017 (Bangkok), and KSS 2018 (Tokyo) have been a successful platform for many scientists and researchers from different countries. During the past 19 years, people interested in knowledge and systems sciences have become a community, and an international academic society has existed for 16 years.

This year KSS was held in Da Nang, Vietnam, during November 29 – December 1, 2019. The conference provided opportunities for presenting interesting new research results and facilitating interdisciplinary discussions, leading to knowledge transfer under the theme of "Knowledge Science in the Age of Big Data." To fit that theme, four distinguished scholars were invited to deliver the keynote speeches:

- Zengru Di (Beijing Normal University, China): "Understanding Science by Network Analysis"
- Takashi Hashimoto (JAIST, Japan): "Emergent Constructive Approach to Language: Computer Simulation, Laboratory Experiment and Brain Measurement"
- Canh Hao Nguyen (Kyoto University, Japan): "Structured Learning for Biomedical Domain"
- Nam Nguyen (IFSR, Austria; Malik Institute, Switzerland and Australia): "Malik Ecopolicy and Sensitivity Model – a cybernetic simulation 'game' for learning about systems thinking and a systems tool for dealing with complex problems"

KSS 2019 received 31 submissions from authors studing and working in China, Japan, New Zealand, Thailand, Vietnam, and the UK, and finally 14 submissions were selected for publication in the proceedings after a double-blind review process. The co-chairs of International Program Committee made the final decision for each submission based on the review reports from the referees, who came from Australia, China, Japan, New Zealand, and Thailand.

For KSS 2019, we received a lot of support and help from many people and organizations. We would like to express our sincere thanks to the authors for their remarkable contributions, all the Technical Program Committee members for their time and expertise in reviewing the papers within a very tight schedule, and the proceedings publisher Springer for their professional help. It is the fourth time that the KSS proceedings are published as a CCIS volume after successful collaboration with Springer during the past three years. We greatly appreciate our four distinguished scholars for accepting our invitation to present keynote speeches at the symposium. Last but not least, we are very indebted to the local organizers for their hard work.

We were happy to the thought-provoking and lively scientific exchanges in the fields of knowledge and systems sciences during the symposium.

November 2019

Jian Chen
Van Nam Huynh
Gia-Nhu Nguyen
Xijin Tang

Organization

Organizer

International Society for Knowledge and Systems Sciences

Host

Duy Tan University, Vietnam

General Chairs

Jian Chen   Tsinghua University, China
Nguyen-Bao Le   Duy Tan University, Vietnam

Program Committee Chairs

Van-Nam Huynh   Japan Advanced Institute of Science and Technology,
                          Japan
Xijin Tang   CAS Academy of Mathematics and Systems Science,
                          China
Jiangning Wu   Dalian University of Technology, China

Technical Program Committee

Quan Bai   University of Tasmania, Australia
Jindong Chen   Beijing Information Science and Technology
                          University, China
Zengru Di   Beijing Normal University, China
Yong Fang   CAS Academy of Mathematics and Systems Science,
                          China
Chonghui Guo   Dalian University of Technology, China
Van-Nam Huynh   Japan Advanced Institute of Science and Technology,
                          Japan
Weihua Li   Auckland University of Technology, New Zealand
Zhenpeng Li   Dali University, China
Jiamou Liu   University of Auckland, New Zealand
Yijun Liu   CAS Institute of Science and Development, China
Mina Ryoke   University of Tsukuba, Japan
Li Song   Shanghai Jiaotong University, China
Bingzhen Sun   Xidian University, China
Leilei Sun   Beihang University, China

Xijin Tang	CAS Academy of Mathematics and Systems Science, China
Cuiping Wei	Yangzhou University, China
Jiang Wu	Wuhan University, China
Jiangning Wu	Dalian University of Technology, China
Nuo Xu	Founder Group and Peking University, China
Yun Xue	South China Normal University, China
Su Yang	Fudan University, China
Thaweesak Yingthawornsuk	King Mongkut's University of Technology Thonburi, Thailand
Wen Zhang	Beijing University of Technology, China
Zhen Zhang	Dalian University of Technology, China

Abstracts of Keynotes

Understanding Science by Network Analysis

Zengru Di

School of Systems Science, Beijing Normal University, Beijing, China
zdi@bnu.edu.cn

Abstract. As many high-quality scientific publication databases, such as the American Physical Society, Scopus, the arXiv, and ISI Web of Knowledge, have become increasingly accessible in recent years, researchers realized that the data should be interpreted from the perspective of complex systems with multiple and evolving interactions between components (e.g. papers, authors, research fields). Using approaches from complex networks and statistical physics, many emergent phenomena have been identified. Examples include the spatial-temporal patterns of researchers' mobility and collaboration, the universal distribution of paper citation across different disciplines, and the collapse of the citation evolution of different papers, and so on. The main contribution of network analysis is to reveal the hidden rules and patterns in scientific research by building the linkage between different scales and dimensions of the system. The related methodologies will be not only valuable for practical use but also will inspire novel ideas and tools for network science.

Emergent Constructive Approach to Language: Computer Simulation, Laboratory Experiment and Brain Measurement

Takashi Hashimoto

Japan Advanced Institute of Science and Technology, Japan
hash@jaist.ac.jp

Abstract. Evolinguitics is an enterprise to clarify the dynamics and mechanisms of origins and evolution of language, thereby, deepening our understanding of humans from an evolutionary perspective. The origin of language is characterized by the biological evolution of abilities related to language and communication, and the evolution of language by the structuralization and complexification of language knowledge as well as communication systems through cultural evolution. In Evolinguistics, two idiosyncrasies of human linguistic communication are the primary focus, namely, using hierarchically organized symbol sequences in language and sharing intentions in symbolic communication. We believe that the integration of these two characteristics made humans co-creative and smart, essentially giving us knowledge co-creation capacity. The emergent constructive approach plays an important role in this research, which is a methodology to analyze complex systems by constructing and operating the evolutionary and emergent process of complex phenomena. In this talk, I will introduce two studies taking this approach. One is an evolutionary simulation of recursive combination which is thought of as the essential ability to form hierarchical structures. The other is a language evolution experiment in a laboratory to consider the process, mechanisms, and a neural basis of symbolic communication systems. After showing findings from these two studies, each related to hierarchy and intention sharing, respectively, I will discuss how these findings can be integrated to understand the evolution of our linguistic communication.

Structured Learning for Biomedical Domain

Canh Hao Nguyen

Kyoto University, Japan
canhhao@kuicr.kyoto-u.ac.jp

Abstract. The biological domain has been blessed with more and more data from biotechnologies as well as data integration tools. In the renaissance of machine learning and artificial intelligence, there is so much promise of data-driven biological knowledge discovery. However, it is not straight forward due to the complexity of the domain knowledge hidden in the data. At any level, be it atom, molecule, cell or organism, there are rich interdependencies among biological components. Machine learning approaches in this domain usually involve analyzing interdependent structures encoded in graphs and related formalisms. In this talk, we will introduce different problems and solutions to the learning problem with structured inputs at different levels. We first will describe learning on graph problems for biological network analysis. The second part will cover the learning problem with structured input data for bio-chemical applications.

Malik Ecopolicy and Sensitivity Model – A Cybernetic Simulation 'game' for Learning About Systems Thinking and a Systems Tool for Dealing with Complex Problems

Nam Nguyen[1,2]

[1] Malik Institute, St Gallen, Switzerland
[2] International Federation for Systems Research (IFSR)
nam.nguyen@mzsg.com

Abstract. Despite many efforts to deal with the various complex issues facing our societies, plans and problem solutions are seldom long lasting, because we, as individuals, and our leaders are most likely to fall into the trap of using traditional linear and separate thinking. It is natural and easy but does not usually deliver long-term solutions in the context of highly complex modern communities and societies. There is an urgent need for innovative ways of thinking and a fresh approach for dealing with the unprecedented and complex challenges facing our world. It is essential for current and future leaders and citizens to be prepared for systems thinking to deal with complex problems in a systemic, integrated, and collaborative fashion; working together to deal with issues holistically, rather than simplistically focusing on isolated/separate features of a system. A revolutionary educational tool (Ecopolicy) has been used as the main mechanism to achieve this aim. Furthermore, the Sensitivity Model ('engine' of Ecopolicy) is used as a systems tool to identify systemic solutions for addressing complex problems in various areas, organisations, businesses, etc. This presentation shares the experience and applications of Ecopolicy in a number of countries (Germany, Australia, Singapore, Vietnam, etc.), as well as the applications of SensiMod in different projects. The presentation concludes with having an interactive, fun, and hands-on Ecopolicy 'play' for the audience.

Keywords: Systems thinking · Systems education · Complexity · Malik Ecopolicy · Cybernetic simulation game · Sensitivity Model

Contents

Estimating the Optimal Number of Clusters in Categorical Data Clustering by Silhouette Coefficient

Duy-Tai Dinh$^{(\boxtimes)}$ ⓘ, Tsutomu Fujinami, and Van-Nam Huynh

School of Knowledge Science, Japan Advanced Institute of Science and Technology,
1-1 Asahidai, Nomi, Ishikawa 923-1292, Japan
{taidinh,fuji,huynh}@jaist.ac.jp

Abstract. The problem of estimating the number of clusters (say k) is one of the major challenges for the partitional clustering. This paper proposes an algorithm named k-SCC to estimate the optimal k in categorical data clustering. For the clustering step, the algorithm uses the kernel density estimation approach to define cluster centers. In addition, it uses an information-theoretic based dissimilarity to measure the distance between centers and objects in each cluster. The silhouette analysis based approach is then used to evaluate the quality of different clusterings obtained in the former step to choose the best k. Comparative experiments were conducted on both synthetic and real datasets to compare the performance of k-SCC with three other algorithms. Experimental results show that k-SCC outperforms the compared algorithms in determining the number of clusters for each dataset.

Keywords: Data mining · Partitional clustering · Categorical data · Silhouette value · Number of clusters

1 Introduction

Clustering is one of the important data mining techniques for discovering knowledge in unstructured multivariate and multidimensional data. The goal of clustering is to identify pattern or groups of similar objects in a given dataset. Each group, or clusters, consists of objects that are similar to one another and dissimilar to objects in other groups. Clustering has been developed in many fields with very diverse applications such as scientific data exploration, information retrieval and text mining, web analysis, marketing, medical diagnostics, computational biology and many others [2]. Clustering algorithms can be classified by type of clustering (hard and soft clustering, flat and hierarchical clustering), type of data (nominal, ordinal, interval scaled and mixed data), by clustering criterion ((probabilistic) model-based and cost-based clustering), or by regime (parametric and non-parametric clustering) [7]. Among these methods, partitional clustering, which is a kind of flat clustering, aims to discover the groupings

© Springer Nature Singapore Pte Ltd. 2019
J. Chen et al. (Eds.): KSS 2019, CCIS 1103, pp. 1–17, 2019.
https://doi.org/10.1007/978-981-15-1209-4_1

present in the data by optimizing a specific objective function and iteratively improving the quality of the partitions [16].

K-means clustering [12] is the most typical representative for the group of partitional clustering algorithms. It is based on the idea of using the cluster centers (means) as representatives of each cluster. K-means has been widely used in many real-life applications due to their simplicity and competitive computational complexity. However, one of the limitations of k-means is that it can not be applied directly to categorical data, which is common in real datasets. To tackle this problem, a data transformation based method can be used to first transform categorical data into a new feature space, and then apply k-means to the newly transformed space to obtain the final results. However, this method has proven to be very ineffective and does not produce good clusters [16]. During the last two decades or so, several attempts have been made to remove the numeric-only limitation of the k-means algorithm and make it applicable to clustering for categorical data [4–6, 8, 9, 13–15, 18]. These algorithms use a similar clustering procedure to k-means, but differ from each other with regards to defining cluster centers and dissimilarity measures for categorical data.

In partitional clustering, the major factors that can impact the performance of this type of algorithms are choosing the initial centroids and estimating the number of clusters [16]. Particularly, in the former case, the random initialization method has been widely used in k-means and k-means-like algorithms for its simplicity. However, this method may yield different clustering results on different runs of the algorithms, and very poor clustering results may occur in some cases. In the latter case, most algorithms assume the number of clusters in advance. However, a fixed number of clusters may lead to difficulties in predicting the actual number of clusters and thus influence the interpretation of the results. Moreover, the over-estimation or under-estimation of the number of clusters will considerably affect the quality of clustering results [10]. Thus, identifying the number of clusters in a dataset is a vital task in clustering analysis. This paper focuses on solving the problem of estimating the number of clusters in categorical data clustering. More specifically, the key contributions of this paper are as follows:

- We propose a k-means-like clustering algorithm based on the silhouette analysis approach to estimate the optimal number of clusters in categorical data clustering, namely k-SCC.
- We perform an extensive experiment on both synthetic and real datasets from the UCI Machine Learning Repository to evaluate the performance of the proposed k-SCC algorithm in terms of clustering quality based on the average silhouette values. The output of the proposed algorithm suggests the optimal number of clusters for each dataset.
- The proposed method is applied to find the optimal number of clusters for a real Sake wine dataset as a case study.

The rest of this paper is organized as follows. Section 2 gives a brief overview of the related work. Section 3 introduces preliminaries and problem statement. Section 4 shows the proposed algorithm for clustering categorical data. Section 5 describes experimental results. Finally, Sect. 6 draws a summary and outlines directions for future work.

2   Related Work

K-means clustering [12] is a well-known algorithm in partitional clustering. It starts by choosing k samples (objects) as the initial centroids. Each sample in the dataset is then assigned into the nearest centroids based on a particular proximity measure, of which the most frequently used are Manhattan, Euclidean and Cosine distances. Once the clusters are formed, the centroids are updated. The algorithm iteratively performs the assignment and update steps until a convergence criterion is met. Given a dataset $D = \{x_1, \ldots, x_n\}$ with n samples and m features, k-means aims to minimize the following objective function:

$$F(U, Z) = \sum_{l=1}^{k} \sum_{i=1}^{n} \sum_{j=1}^{m} u_{il} \times d(x_{ij}, z_{lj}) \tag{1}$$

where k be the number of clusters; $U = [u_{il}]$ is an $n \times k$ partition matrix that satisfies $u_{il} \in \{0, 1\}$ and $\sum_{l=1}^{k} u_{il} = 1$ $(1 \leq i \leq n; 1 \leq l \leq k)$; $Z = \{Z_l, l = 1, \ldots, k\}$ is a set of cluster centers in which Z_l consists of m values $(z_1^l, z_2^l, \ldots, z_m^l)$, each being the mean of an attribute j in cluster Z_l and defined as $z_j^l = \frac{\sum_{x_i \in Z_l} x_{ij}}{|Z_l|}$; $d(\cdot, \cdot)$ is the squared Euclidean between two attribute values. K-means has many advantages such as easy implementation, easy interpretation and efficient computation. However, working only on numerical data prohibits some applications of k-means. Particularly, it can not be directly applied to categorical data which is fairly common in real datasets.

To tackle this problem, several k-means-like algorithms have been proposed for categorical data clustering. K-modes [8,9] extends k-means by using *modes* to define cluster centers and the simple matching dissimilarity measure to measure the distance of categorical objects. Let x_a and x_b be two categorical values. The simple matching distance between x_a and x_b is given by:

$$\delta(x_a, x_b) = \begin{cases} 0 & \text{if } x_a = x_b \\ 1 & \text{if } x_a \neq x_b \end{cases} \tag{2}$$

As k-means, k-modes is also an optimization problem and has similar runtime, benefits, and disadvantages [9].

K-representatives [18] applies the other approach to define cluster representative. It uses the distribution of categorical values appearing in each attribute to define the representative at that attribute. The dissimilarity between each object and representative is measured based on the multiplication of relative frequencies of categorical values within the cluster and the simple matching measure between categorical values. K-centers [4] uses the kernel density estimation method to define the center of a categorical cluster, called the probabilistic center. It incorporates a built-in feature weighting in which each attribute is automatically assigned with a weight to measure its contribution for the clusters. More recently, several works have been proposed to improve clustering results or deal with the problem of missing values in categorical data [5,6,13,15]. In [15], three modifications of k-representatives were introduced. The first version (Modified-1) replaces the simple matching measure with an information-theoretic based dissimilarity measure. The second version (Modified-2) uses the new dissimilarity measure with the concept of cluster centers proposed in k-centers [4] to form clusters. The third version (Modified-3) uses the dissimilarity as in Modified-2 and incorporates with a modified representative of cluster centers using the kernel density estimation method. Modified-3 was shown to be more efficient than the former two versions in terms of clustering quality.

The silhouette coefficient [17] is an internal measure for cluster validation. It considers both the intra-cluster and inter-cluster distances. For each data sample x_i, the average of the distances to all samples in the same cluster is first calculated and set to $a(x_i)$. For each cluster that does not contain this sample, the average distance of x_i to all samples in each cluster is calculated. Then the smallest of these distances is taken and set to $b(x_i)$. The two values $a(x_i)$ and $b(x_i)$ are used to estimate the silhouette coefficient of x_i. The average of all the silhouettes in the dataset is called the average silhouette for all samples in the dataset. The average silhouette can be used to evaluate the quality of a clustering. In addition, the optimal cluster number can be determined by maximizing the value of this index. Several works based on silhouette coefficient have been proposed to estimate the number of clusters in numerical data clustering [1,20]. To the best of our knowledge, no prior work has been conducted on estimating the number of clusters in categorical data clustering using the silhouette coefficient. The next section shows preliminary definitions of the proposed framework.

3   Preliminaries

Let n and m be the number of objects and attributes in a categorical dataset, respectively. A categorical dataset D is an $n \times m$ matrix ($n \gg m$) in which each element at position (i, j) ($1 \le i \le n$, $1 \le j \le m$) stores the value of an object x_i at the jth attribute. A categorical object $x_i \in D$ ($1 \le i \le n$) is a tuple of m values $x_i = (x_{i1}, x_{i2}, \ldots, x_{im}) \in A_1 \times A_2 \times \cdots \times A_m$, where A_j is a categorical attribute which is characterized by a finite domain $\mathcal{O}_j$ such that $\mathrm{DOM}(A_j) = |\mathcal{O}_j|$ (> 1) discrete values. In addition, a category in $\mathcal{O}_j$ is denoted by o_{ij} ($1 \le i \le |\mathcal{O}_j|$). For example, Table 1 shows a categorical dataset that contains ten objects with six categorical attributes.

Table 1. A categorical dataset

Obj	Attr					
	A_1	A_2	A_3	A_4	A_5	A_6
x_1	a_1	d_2	b_3	e_4	a_5	c_6
x_2	d_1	a_2	a_3	b_4	c_5	a_6
x_3	d_1	d_2	d_3	c_4	c_5	a_6
x_4	b_1	e_2	c_3	e_4	a_5	c_6
x_5	a_1	d_2	a_3	a_4	a_5	e_6
x_6	a_1	a_2	c_3	e_4	a_5	c_6
x_7	b_1	e_2	e_3	e_4	a_5	c_6
x_8	d_1	c_2	d_3	e_4	a_5	c_6
x_9	d_1	c_2	d_3	e_4	a_5	c_6
x_{10}	d_1	d_2	b_3	b_4	c_5	a_6

Definition 1 (Clusters). Given a categorical dataset $D = \{x_1, \ldots, x_n\}$, let $C = \{C_1, C_2, \ldots, C_k\}$ be a set of k disjoint subsets that contain the indices of objects in D. These subsets are called clusters if they satisfy the two conditions: $C_l \cap C_{l'} = \emptyset$ for all $l \neq l'$ and $\bigcup_{l=1}^{k} C_l = D$. The number of objects in cluster C_l is denoted by n_l.

Definition 2 (Relative frequency). Let there be a cluster C_l, the relative frequency of a category o_{ij}^l $(1 \leq i \leq n_l, 1 \leq j \leq m)$ occurring in C_l at the jth attribute is defined as:

$$f_l(o_{ij}^l) = \frac{\#_l(o_{ij}^l)}{n_l} \tag{3}$$

with $\#_l(o_{ij}^l)$ being the number of o_{ij}^l appearing in cluster C_l at the jth attribute.

The relative frequency of a category o_{ij} occurring at the jth attribute of dataset D is defined as:

$$f(o_{ij}) = \frac{\#(o_{ij})}{n} \tag{4}$$

and $\#(o_{ij})$ denotes the number of o_{ij} occurring in dataset D at the jth attribute.

In partitional clustering, each cluster is represented by its center. From a statistical perspective, the cluster center of a numeric cluster is the expectation of a continuous random variable associated with the data, based on the assumption that the variable follows a Gaussian distribution [4]. Following this perspective, the center of a categorical cluster can be estimated by using the kernel density estimation method, called the probabilistic center [4–6,13,15]. In particular, the center of a categorical cluster C_l is denoted as $Z_l = \{z_j^l\}_{j=1}^{m}$ where the jth element being a vector in the probability space and $\mathcal{O}_j^l$, P_j^l (in Definition 5) serves as the sample space and the probability measure defined on the Borel set of the sample space with regard to data subset C_l, respectively.

Definition 3 (Kernel density estimation method). Let there be a cluster C_l, let X_j^l be the random variable associated with observations x_i ($1 \leq i \leq n_l$) occurring in cluster C_l at the jth attribute and $p(X_j^l)$ be its probability of density. Let $\mathcal{O}_j^l$ denote the set of categories occurring at the jth attribute of C_l such that $\mathcal{O}_j^l = \bigcup_{i=1}^{n_l} x_{ij}$ and $\lambda_l \in [0, 1]$ be the unique smoothing bandwidth for cluster C_l. For each value o_{ij}^l in $\mathcal{O}_j^l$ ($1 \leq i \leq n_l$), the variation on Aitchison & Aitken's kernel function is denoted and defined as:

$$K(X_j^l, o_{ij}^l, \lambda_l) = \begin{cases} 1 - \frac{|\mathcal{O}_j^l|-1}{|\mathcal{O}_j^l|}\lambda_l & \text{if } X_j^l = o_{ij}^l \\ \frac{1}{|\mathcal{O}_j^l|}\lambda_l & \text{otherwise} \end{cases} \tag{5}$$

Note that the above kernel function is estimated in terms of the cardinality of the subdomain $\mathcal{O}_j^l$ of cluster C_j. Let $\hat{p}(X_j^l, \lambda_l, C_l)$ be the kernel estimator of $p(X_j^l)$. It is defined as:

$$\hat{p}(X_j^l, \lambda_l, C_l) = \sum_{o_{ij}^l \in \mathcal{O}_j^l} f_l(o_{ij}^l)K(X_j^l, o_{ij}^l, \lambda_l) \tag{6}$$

Definition 4 (Smoothing bandwidth parameter). Let there be a cluster C_l, a smoothing bandwidth parameter using the least square cross-validation is used to minimize the total error of the resulting estimation over data objects in this cluster. The optimal smoothing parameter for C_l is denoted and defined as:

$$\lambda_l = \frac{1}{(n_l - 1)} \frac{\sum_{j=1}^m (1 - \sum_{o_{ij}^l \in \mathcal{O}_j^l} [f_l(o_{ij}^l)]^2)}{\sum_{j=1}^m (\sum_{o_{ij}^l \in \mathcal{O}_j^l} [f_l(o_{ij}^l)]^2 - \frac{1}{|\mathcal{O}_j^l|})} \tag{7}$$

Definition 5 (Probabilistic Center of a Categorical cluster). Let there be a cluster $C_l = \{x_1, x_2, \ldots, x_{n_l}\}$ where $x_i = (x_{i1}, x_{i2}, \ldots, x_{im})$ ($1 \leq i \leq n_l$). Let $\mathcal{O}_j^l$ be a set of categories occurring at the jth attribute in cluster C_l. The center of C_l is denoted and defined as:

$$Z_l = \{z_1^l, z_2^l, \ldots, z_m^l\} \tag{8}$$

where the value at j^{th} element of Z_l is a probability distribution on $\mathcal{O}_j^l$ estimated by a kernel density estimation method using Eq. (6) and is defined as:

$$z_j^l = [P_j^l(o_{1j}^l), P_j^l(o_{2j}^l), \ldots, P_j^l(o_{|\mathcal{O}_j^l|j}^l)] \tag{9}$$

and the value of each category o_{ij}^l ($1 \leq i \leq |\mathcal{O}_j^l|$) is measured by using Eqs. (3), (5) and (6) as follows:

$$P_j^l(o_{ij}^l) = \begin{cases} \lambda_l \frac{1}{|\mathcal{O}_j^l|} + (1 - \lambda_l)f_l(o_{ij}^l) & \text{if } o_{ij}^l \in \mathcal{O}_j^l \\ 0 & \text{otherwise} \end{cases} \tag{10}$$

The probabilistic estimation of a category shown in Eq. (10) can be seen as a Bayes-type probability estimator where the uniform probability ($\frac{1}{|O_{j}^{l}|}$) is as a prior and the frequency estimator ($f_l(o_{ij}^l)$) is as the posterior [4].

Methods for quantifying the distance between categorical objects and cluster centers have been extensively studied in recent years [3, 8, 11, 15, 19]. In general, similarity measures which allow the comparison of values in an attribute can be classified into three types. The first type assigns possible values that are different than zero for the similarity in which a match occurs, while a value of zero is assigned for the similarity in which a value of mismatch occurs. The second type assigns a value of one for the similarity in which a value match occurs, while possible value different than one is assigned for the similarity in which a value of mismatch occurs. The last type defines different values when matching and mismatching occur [19]. The Lin similarity [11] falls into the last type. It defines similarity as the relationship between the common and different information component from the view of information theory, which uses the logarithmic function to calculate the real value in such a way that less frequent words have a higher information gain. Based on Lin similarity, an information-theoretic based dissimilarity measure has been proposed for categorical data [3, 15], which is used in this paper to determine the distance between a categorical object and its cluster center.

Definition 6 (Dissimilarity between two categories). The similarity of two categories o_{ij} and $o_{i'j}$ occurring in two objects x_i and $x_{i'}$ at the jth attribute is defined as:

$$\text{sim}_j(o_{ij}, o_{i'j}) = \frac{2\log f(o_{ij}, o_{i'j})}{\log f(o_{ij}) + \log f(o_{i'j})} \tag{11}$$

with $f(o_{ij}, o_{i'j}) = \frac{\#(o_{ij}, o_{i'j})}{|D|}$ be the relative frequency of two categorical objects in dataset D that receive the value belonging to $\{o_{ij}, o_{i'j}\}$ at the jth attribute. The dissimilarity between o_{ij} and $o_{i'j}$ at the jth attribute is measured as:

$$\text{dsim}_j(o_{ij}, o_{i'j}) = 1 - \text{sim}_j(o_{ij}, o_{i'j}) = 1 - \frac{2\log f(o_{ij}, o_{i'j})}{\log f(o_{ij}) + \log f(o_{i'j})} \tag{12}$$

Definition 7 (Dissimilarity between two categorical objects). Let there be two categorical objects $x_i = (x_{i1}, x_{i2}, \ldots, x_{im})$ and $x_{i'} = (x_{i'1}, x_{i'2}, \ldots, x_{i'm})$, the dissimilarity of x_i and $x_{i'}$ is defined as:

$$\text{dsim}(x_i, x_{i'}) = \sum_{j=1}^{m} \text{dsim}_j(x_{ij}, x_{i'j}) \tag{13}$$

Table 2 shows the pairwise distance matrix of all data objects shown in Table 1. This is a symmetric matrix where $\text{dsim}(x_i, x_{i'}) = \text{dsim}(x_{i'}, x_i)$ and $\text{dsim}(x_i, x_i) = 0$ $(1 \leq i, i' \leq n)$. Intuitively, the dissimilarity of two objects is zero if they are identical. The upper bound dissimilarity value of them is exactly the number of attributes, where categorical values at each attribute of two objects are different.

Table 2. Dissimilarity matrix of categorical objects

	1	2	3	4	5	6	7	8	9	10
1	0.0000	4.3518	3.7898	1.4107	2.308	1.4107	1.3645	1.7617	1.7617	3.3256
2	4.3518	0.0000	1.4872	4.2182	3.1476	3.7875	4.172	3.6046	3.6046	1.0262
3	3.7898	1.4872	0.0000	4.4165	2.9759	4.4165	4.3866	3.2191	3.2191	0.8917
4	1.4107	4.2182	4.4165	0.0000	2.9497	0.8614	0.3845	1.6281	1.6281	4.383
5	2.308	3.1476	2.9759	2.9497	0.0000	2.5191	2.9035	3.2858	3.2858	2.9827
6	1.4107	3.7875	4.4165	0.8614	2.5191	0.0000	1.2458	1.6281	1.6281	4.383
7	1.3645	4.172	4.3866	0.3845	2.9035	1.2458	0.0000	1.5983	1.5983	4.3368
8	1.7617	3.6046	3.2191	1.6281	3.2858	1.6281	1.5983	0.0000	0.0000	3.7694
9	1.7617	3.6046	3.2191	1.6281	3.2858	1.6281	1.5983	0.0000	0.0000	3.7694
10	3.3256	1.0262	0.8917	4.383	2.9827	4.383	4.3368	3.7694	3.7694	0.0000

Definition 8 (Distance between an object and cluster center). Let there be a cluster C_l with its center $Z_l = \{z_1^l, z_2^l, \ldots, z_m^l\}$ and a categorical object $x_i = (x_{i1}, x_{i2}, \ldots, x_{im})$. Let $\mathcal{O}_j^l$ be a set of categories appearing at the jth attribute of z_j^l. The dissimilarity between x_i and Z_l at the jth attribute is measured by accumulating the probability distribution on $\mathcal{O}_j^l$ and the dissimilarity between j^{th} component x_{ij} of the object x_i and the jth component z_j^l of the center Z_l, which is formulated as:

$$\text{dis}_j(x_i, Z_l) = \sum_{o_{ij}^l \in \mathcal{O}_j^l} P_j^l(o_{ij}^l)\text{dsim}_j(x_{ij}, o_{ij}^l) \tag{14}$$

The distance between x_i and cluster center Z_l is then measured as:

$$\text{dis}(x_i, Z_l) = \sum_{j=1}^{m} \text{dis}_j(x_i, Z_l) \tag{15}$$

Based on the above distance measure, the categorical data clustering algorithm aims to minimize the following optimization function:

$$F(U, Z) = \sum_{l=1}^{k} \sum_{i=1}^{n} u_{i,l} \times dis(x_i, Z_l) \tag{16}$$

subject to

$$\begin{cases} \sum_{l=1}^{k} u_{i,l} = 1 & 1 \leq i \leq n \\ u_{i,l} \in \{0,1\} & 1 \leq l \leq k, \ 1 \leq i \leq n \end{cases} \tag{17}$$

where $U = [u_{i,l}]_{n \times k}$ is the partition matrix in which $u_{i,l}$ takes value 1 if object x_i is in cluster C_l and 0 otherwise.

Definition 9 (Silhouette value of a categorical object). Let there be a cluster C_l, let x_i be a categorical object in C_l $(1 \leq i \leq n_l)$. Let $a(x_i)$ be the average distance of the x_i to all other members of the same cluster C_l. Let $C_{l'}$ be some cluster other than C_l and let $d(x_i, C_{l'})$ be the average distance of the x_i to all members of $C_{l'}$. Compute $d(x_i, C_{l'})$ for all clusters $C_{l'}$ other than C_l and let $b(x_i) = \min_{C_{l'} \neq C_l} d(x_i, C_{l'})$. If cluster C_α ($1 \leq \alpha \leq k$) satisfies the condition that $b(x_i) = d(x_i, C_\alpha)$ then C_α is called the neighbor of x_i and is considered as the second-best cluster for the x_i. The silhouette value of x_i is denoted and defined as:

$$s(x_i) = \frac{b(x_i) - a(x_i)}{\max\{a(x_i), b(x_i)\}} \tag{18}$$

The Eq. (18) shows that the silhouette value is between -1 and 1. A large positive value of $s(x_i)$, i.e. $a(x_i)$ approximately equals to zero, indicates that the *within* dissimilarity $a(x_i)$ is much smaller than the smallest *between* dissimilarity $b(x_i)$ and thus x_i is well-clustered. A large negative value of $s(x_i)$, i.e. $b(x_i)$ approximately equals to zero, indicates that x_i is poor-clustered. If $s(x_i)$ is about zero, i.e. $a(x_i)$ approximately equals to $b(x_i)$, indicate that x_i lies between two clusters. In general, $s(x_i)$ measures how well object x_i has been classified into cluster C_l.

Definition 10 (Average silhouette value). Let there be a set of k clusters $C = \{C_1, C_2, \ldots, C_k\}$, the average of all silhouette values for all categorical objects in the dataset is called the *average silhouette value* and is defined as:

$$s_k = \frac{\sum_{i=1}^{n} s(x_i)}{n} \tag{19}$$

In this paper, the problem of estimating the number of clusters in categorical data clustering is to find the optimal k to maximize the average silhouette value shown in Eq. (19).

The next section proposes an algorithm named k-SCC for estimating the number of clusters in categorical data clustering.

4   The Proposed k-SCC Algorithm

The proposed k-SCC algorithm is based on the general framework depicted in Fig. 1. According to this model, the proposed algorithm partitions a categorical data into k groups and then computes the average silhouette value for the current iteration. This process works in the same manner for every index k in the range of predefined minimum and maximum number of clusters. Finally, the index that yields the largest value of the average silhouette is selected as the optimal number of clusters for that categorical data.

The pseudo code of the k-SCC algorithm is shown in Algorithm 1. The input of this algorithm is a categorical data D and two predefined minimum and maximum numbers of clusters, denoted by $k_{\min}$ and $k_{\max}$, respectively. By default, $k_{\min}$ and $k_{\max}$ are in range $[2, n-1]$, where n is the number of objects in the given dataset. A set of average silhouettes, namely $SilSet$, is used to keep the average silhouette at every iteration of the algorithm (line 1). The algorithm then computes the dissimilarity matrix that contains dissimilarity of pairs of

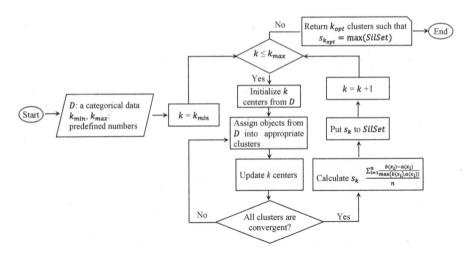

Fig. 1. The flowchart of k-SCC algorithm

Algorithm 1. THE k-SSC ALGORITHM

---

input : D: a categorical dataset, $k_{\min}$, $k_{\max}$: the minimum and maximum value of k.

output: k clusters of D

1 SilSet $\leftarrow \emptyset$, $k = k_{\min}$
2 Compute the dissimilarity matrix of all categorical objects in D using Eq. (13)
3 **while** $k \leq k_{\max}$ **do**
4 $\quad$ Initialize k cluster centers $Z^{(0)} = \{Z_1^{(0)}, \ldots, Z_k^{(0)}\}$
5 $\quad$ $U \leftarrow \emptyset$, $t = 0$
6 $\quad$ **while** Partitions are not convergent **do**
7 $\quad\quad$ Keep $Z^{(t)}$ fixed, generate $U^{(t)}$ to minimize the distances between objects and cluster centers using Eq. (15)
8 $\quad\quad$ Keep $U^{(t)}$ fixed, update $Z^{(t)}$ using Eq. (8)
9 $\quad\quad$ $t = t + 1$
10 $\quad$ Compute the average silhouette value s_k using Eq. (19)
11 $\quad$ SilSet $\leftarrow s_k$, $k = k + 1$
12 **return** k_{opt} clusters such that $s_{k_{opt}} = \max(SilSet)$;

---

objects in D based on the Eq. (13) (line 2). This matrix is used later for computing the *within* and *between* dissimilarities in average silhouette values. For every index k in the range of $k_{\min}$ and $k_{\max}$, k-SCC performs assignment and update steps to partition categorical objects into k clusters (lines 4 to 9). In the initial step, k-SCC initiates k cluster centers by randomly selecting k objects from D, one for each cluster (line 4). In the next step, k-SCC assigns each object in D to its nearest cluster based on the dissimilarity function shown in Eq. (15). It then updates k cluster center based on Eq. (8). The algorithm iterates the assignment and update steps until no further reassignment of objects takes place. The indexes of objects in k clusters and the proximity matrix are used to compute the average silhouette value based on the Eq. (19) (line 10). The obtained average silhouette value is then put into the $SilSet$ (line 11). The k-SCC works in the same manner for every index k until k is larger than $k_{\max}$ (lines 3 to 11). Finally, the index k that has the largest value of average silhouette in $SilSet$ is selected as the optimal index k, namely k_{opt}, for the given dataset. The algorithm then returns k_{opt} clusters as the output of the algorithm (line 12).

5   Comparative Experiment

In this section, the proposed algorithm was tested on both real and synthetic datasets. The Car evaluation, Chess, Connect-4, Nursery, Soybean (small), Spect heart and Tictactoe are real datasets from the UCI repository[1]. The SD5K and SD10K are synthetic datasets generated by the Dataset Generator[2]. The characteristics of these datasets are shown in Table 3. The performance of k-SCC is compared with other three clustering algorithms: k-modes [8], Modified-3 [15] and an additional version of the k-SCC that uses the simple matching dissimilarity measure to compute the dissimilarity matrix in Algorithm 1, namely k-SCC+. All algorithms were implemented in Python using PyCharm. Experiments were performed on a high-performance VPCC cluster[3] equipped with an Intel Xeon Gold 6130 2.1 GHz (16 Cores×2), 64 GB of RAM, running CentOS 7.2 for each CPU node. The source code of k-SCC and datasets are provided at http://bit.ly/2HiwgKl. Since the four algorithms use random initialization to initially form centers of clusters, it may lead to non-repeatable clustering results. For that reason, each of the four algorithms was run 100 times for the initialization step and the best one was selected for each dataset.

Table 3. Characteristics of the experimental datasets

Dataset	#instances	#attributes	#classes	Type
Car Evaluation	1,728	6	4	Real-life
Chess	3,196	36	2	Real-life
Connect-4	10,000	42	3	Real-life
Nursery	12,960	8	5	Real-life
Soybean (small)	47	35	4	Real-life
Spect Heart	267	22	2	Real-life
Tic-Tac-Toe	958	9	2	Real-life
SD5K	5,000	6	4	Synthetic
SD10K	10,000	6	2	Synthetic

---

[1] https://archive.ics.uci.edu/ml/index.php.
[2] http://www.datasetgenerator.com.
[3] https://www.jaist.ac.jp/iscenter/en/mpc/vpcc/.

5.1   Experimental Results

In the experiment, the average silhouette values were measured for the various number of clusters on each dataset. More specifically, k is varied between two and ten for each dataset, then choose the best k by comparing the average silhouette results obtained for the different k values. Results are shown in Fig. 2. In this figure, vertical axes denote the average silhouette value, and horizontal axes indicate the number of clusters used. In general, k-SCC has higher average silhouette values than those of k-modes, Modified-3 and k-SCC+ for all datasets. On Car Evaluation, it can be classified into two or four clusters, which correspond to the largest and second largest average silhouettes for this dataset. On Chess, this dataset can be classified into two or three clusters based on its largest and second largest average silhouettes, respectively. Similar results can be observed for the other datasets. The results show that the average silhouette values of k-SCC and k-SCC+ are higher than those of k-modes. It means that the kernel density estimation used to form cluster centers used in k-SCC and k-SCC+ are more efficient than the modes used in k-modes. In addition, the information-theoretic based dissimilarity measure used in k-SCC and k-SCC+ are more efficient than the simple matching dissimilarity used in k-modes. It can be also observed that the average silhouettes of k-SCC are better than those of k-SCC+. Thus, the information-theoretic based dissimilarity measure used to compute the dissimilarity matrix is more effective than the simple matching used in k-SCC+. Furthermore, the average silhouette of k-SCC are better than those of the Modified-3 in most cases. It is worth to note that the runtime of Modified-3 is much higher than that of k-SCC because the former uses a feature weighting scheme to automatically measure the contribution of individual attributes for the cluster. Thus, k-SCC is more efficient in both estimation and computation than Modified-3. In general, the proposed framework is sensitive to the choice of dissimilarity measure to determine distances between data objects and the way to perform the clustering step.

5.2   A Case Study: Sake Wine Dataset

Another experiment was performed to evaluate the performance of compared algorithms on the Sake wine dataset, which is a real dataset collected from Fujinami lab[4]. More specifically, the taste sensing system named TS-5000Z, which is employed the same mechanism as that of the human tongue is used to convert the taste of Sake wine into numerical data. The TS-5000Z evaluates two types of taste: initial taste, which is the taste perceived when food first enters the mouth, and aftertaste, which is the persistent taste that remains in the mouth after the food has been swallowed. The initial taste is indicated by six features: *sourness, bitterness, astringency, umami, saltiness* and *sweetness*, while the aftertaste is indicated by *aftertaste from astringency* and *richness*. The dataset contains 68 Sake samples with eight features corresponding to the initial taste and aftertaste.

---

[4] http://www.jaist.ac.jp/~fuji/index.html.

Figure 3 shows a classification of the Sake data by using the complete-linkage hierarchical clustering. If the dendrogram is cut at the height of eight, nine or ten, then four (with one outlier), three or two clusters are obtained, respectively. The original Sake wine dataset obtained by using the TS-5000Z is a numeric data. We first pre-process the original data so as to be applicable to k-SCC. In this work, Kansei words with five linguistic grades: *very low, low, neither, high, very high* are used to convert the numeric values into categories in each feature. We first discretized each attribute to five scales corresponding to five linguistic

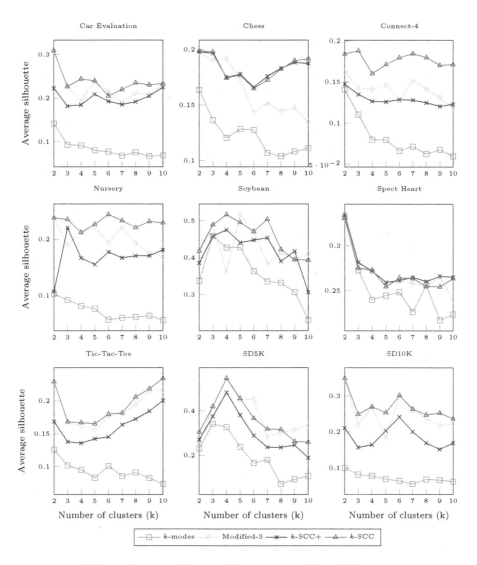

Fig. 2. Average silhouette for various number of clusters

grades. We then mapped each value in each attribute into the corresponding grade. The average silhouette values of k-SCC and compared algorithms are shown in Table 4. It is observed that the k-SCC outperforms k-modes, Modified-3 and k-SCC+ in most cases, except for $k = 10$ where the average silhouette of k-SCC+ is higher than one of k-SCC. Figure 4 shows silhouette plots of the Sake wine data for $k \in [2, 10]$. According to the results in Table 4, the Sake samples can be classified into three or four clusters, which correspond to the largest and second largest average silhouettes for this dataset. In other words, the recommended number of clusters match the results obtained in Fig. 3.

6   Summary and Discussion

This paper has proposed an algorithm named k-SCC for estimating the optimal numbers of clusters in categorical data clustering. K-SCC uses the silhouette approach to find the optimal k corresponding to the highest average silhouette value in each dataset. To perform the clustering steps, the proposed algorithm uses the kernel density estimation method and an information-theoretic based dissimilarity to form cluster centers and measure the distance between objects and cluster centers, respectively. This dissimilarity is also used to build the dissimilarity matrix when computing the average silhouette values. The experimental results have shown that the proposed algorithm outperforms the other three compared algorithms on both synthesis and real datasets.

The advantages of the proposed algorithm are as follows. First, this method can deal with categorical data, which is very popular in many real-life applications. In addition, the proposed method can be extended to estimate the number

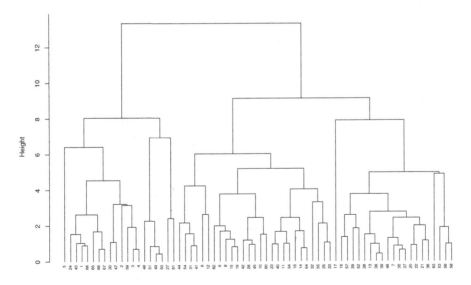

Fig. 3. Hierarchical clustering on Sake wine data using Complete-lingkage

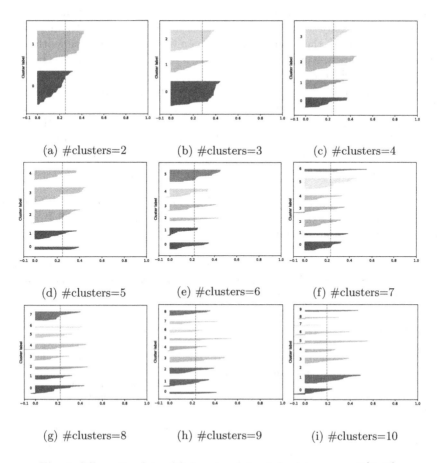

(a) #clusters=2   (b) #clusters=3   (c) #clusters=4

(d) #clusters=5   (e) #clusters=6   (f) #clusters=7

(g) #clusters=8   (h) #clusters=9   (i) #clusters=10

Fig. 4. Silhouette plots of Sake wine dataset for k in range of $[2, 10]$

Table 4. Average silhouette values on the Sake wine dataset

k	Algos			
	k-modes	Modified-3	k-SCC+	k-SCC
2	0.1461	**0.2501**	0.2014	**0.2501**
3	0.1405	0.2744	0.2096	**0.2811**
4	0.1562	0.2252	0.1584	**0.2585**
5	0.1351	**0.2511**	0.1923	0.2451
6	0.1736	0.2135	0.1879	**0.2200**
7	0.1778	**0.2556**	0.2128	0.2328
8	0.1512	0.2028	0.2204	**0.2217**
9	0.1168	0.2168	0.1968	**0.2288**
10	0.1412	0.1469	**0.2148**	0.1924

of clusters in mixed numeric and categorical data clustering. Second, the proposed method is a partitional clustering, which means that it will terminate at a local optimum. Third, the proposed method can be upgraded by using other dissimilarity measures to compute the dissimilarity matrix and distances between data objects and cluster centers.

The inherent limitations of the proposed algorithm are as follows. The average silhouette values strongly depend on the dissimilarity measure used to compute the pairwise dissimilarity matrix and the clustering scheme to partition the input datasets. In addition, the proposed method has high costs when applying for large-scale datasets. Thus, using a suitable clustering framework to get high silhouette values and reducing the computational complexity is necessary for this task.

In future work, we will extend the proposed approach to the problem of estimating the number of clusters in mixed data and design a parallel method to speed up the computational process.

Acknowledgment. This paper is based upon work supported in part by the Air Force Office of Scientific Research/Asian Office of Aerospace Research and Development (AFOSR/AOARD) under award number FA2386-17-1-4046.

References

1. Azimi, R., Ghayekhloo, M., Ghofrani, M., Sajedi, H.: A novel clustering algorithm based on data transformation approaches. Expert Syst. Appl. **76**, 59–70 (2017)
2. Berkhin, P.: A survey of clustering data mining techniques. In: Kogan, J., Nicholas, C., Teboulle, M. (eds.) Grouping Multidimensional Data, pp. 25–71. Springer, Heidelberg (2006). https://doi.org/10.1007/3-540-28349-8_2
3. Boriah, S., Chandola, V., Kumar, V.: Similarity measures for categorical data: a comparative evaluation. In: Proceedings of the 2008 SIAM International Conference on Data Mining, pp. 243–254. SIAM (2008)
4. Chen, L., Wang, S.: Central clustering of categorical data with automated feature weighting. In: IJCAI, pp. 1260–1266 (2013)
5. Dinh, D.-T., Huynh, V.-N.: k-CCM: a center-based algorithm for clustering categorical data with missing values. In: Torra, V., Narukawa, Y., Aguiló, I., González-Hidalgo, M. (eds.) MDAI 2018. LNCS (LNAI), vol. 11144, pp. 267–279. Springer, Cham (2018). https://doi.org/10.1007/978-3-030-00202-2_22
6. Dinh, D.T., Huynh, V.N., Sriboonchita, S.: Data for: clustering mixed numeric and categorical data with missing values (2019)
7. Hennig, C., Meila, M., Murtagh, F., Rocci, R.: Handbook of Cluster Analysis. Chapman & Hall/CRC Handbooks of Modern Statistical Methods. CRC Press, Boca Raton (2015)
8. Huang, Z.: Clustering large data sets with mixed numeric and categorical values. In: Proceedings of the First Pacific Asia Knowledge Discovery and Data Mining Conference, pp. 21–34. World Scientific, Singapore (1997)
9. Huang, Z.: Extensions to the k-means algorithm for clustering large data sets with categorical values. Data Min. Knowl. Discov. **2**(3), 283–304 (1998)
10. Liang, J., Zhao, X., Li, D., Cao, F., Dang, C.: Determining the number of clusters using information entropy for mixed data. Pattern Recogn. **45**(6), 2251–2265 (2012)

11. Lin, D.: An information-theoretic definition of similarity. In: Proceedings of the Fifteenth International Conference on Machine Learning, pp. 296–304 (1998)
12. MacQueen, J.: Some methods for classification and analysis of multivariate observations. In: Proceedings of the Fifth Berkeley Symposium On Mathematical Statistics and Probability, Oakland, CA, USA, vol. 1, pp. 281–297 (1967)
13. Nguyen, T.-P., Dinh, D.-T., Huynh, V.-N.: A new context-based clustering framework for categorical data. In: Geng, X., Kang, B.-H. (eds.) PRICAI 2018. LNCS (LNAI), vol. 11012, pp. 697–709. Springer, Cham (2018). https://doi.org/10.1007/978-3-319-97304-3_53
14. Nguyen, T.H.T., Dinh, D.T., Sriboonchitta, S., Huynh, V.N.: A method for k-means-like clustering of categorical data. J. Ambient. Intell. Hum. Comput. 1–11 (2019). https://doi.org/10.1007/s12652-019-01445-5
15. Nguyen, T.-H.T., Huynh, V.-N.: A k-means-like algorithm for clustering categorical data using an information theoretic-based dissimilarity measure. In: Gyssens, M., Simari, G. (eds.) FoIKS 2016. LNCS, vol. 9616, pp. 115–130. Springer, Cham (2016). https://doi.org/10.1007/978-3-319-30024-5_7
16. Reddy, C.K., Vinzamuri, B.: A survey of partitional and hierarchical clustering algorithms. In: Data Clustering: Algorithms and Applications, pp. 87–110. Chapman and Hall/CRC (2013)
17. Rousseeuw, P.J.: Silhouettes: a graphical aid to the interpretation and validation of cluster analysis. J. Comput. Appl. Math. **20**, 53–65 (1987)
18. San, O.M., Huynh, V.N., Nakamori, Y.: An alternative extension of the k-means algorithm for clustering categorical data. Int. J. Appl. Math. Comput. Sci. **14**, 241–247 (2004)
19. dos Santos, T.R., Zárate, L.E.: Categorical data clustering: what similarity measure to recommend? Expert. Syst. Appl. **42**(3), 1247–1260 (2015)
20. Ünlü, R., Xanthopoulos, P.: Estimating the number of clusters in a dataset via consensus clustering. Expert. Syst. Appl. **125**, 33–39 (2019)

An Improved Two-Stage Multi-person Pose Estimation Model

Sutong Wang[(✉)] [iD], Yanzhang Wang, Xuehua Wang, Xin Ye,
Huaiming Li, and Xuelong Chen

Institution of Information and Decision Technology,
Dalian University of Technology, Dalian 116023, China
sutongwang@mail.dlut.edu.cn

Abstract. Generally, multi-person pose estimation plays a crucial role in behavior recognition in images and videos. Previously, pose estimation of a single person is popular and achieves high prediction accuracy with the development of deep learning. However, pose estimation of multi-person remains to be a huge challenge and cannot achieve the same effect as that of a single person. It mainly results from the rare, missing or incorrect location detection and overlap of pose, which are usually caused by incomplete person identification. Therefore, we propose an improved two-stage multi-person pose estimation model (ITMPE) to further improve the performance of multi-person pose estimation. The first stage, Mask R-CNN is used for person identification. The second stage, processed images or videos with identified people only are fed into OpenPose model for multi-person pose estimation. The comparative experiments show that our proposed model achieves a significant improvement than original model. Our proposed model reduces the MSE, MAE by around 27.38%, 21.57% and increases R^2, Mean values by 49.80% and 96.91% on average, respectively. The improvement in person identification and misclassification are shown in our comparison images. More people are captured and given the pose estimation, which directly affect the performance of behavior recognition.

Keywords: Pose estimation · Multi-person · Instance segmentation · Complex scenarios

1 Introduction

In general, multi-person pose estimation plays a crucial role in behavior recognition in images and videos. In advance, single person pose estimation now achieves high prediction accuracy with the support of deep learning. In contrast, it is a unique challenge to estimate multi-person postures in images or videos. First, human interactions can cause complex spatial disturbances caused by contact, occlusion, or joint of the limb, which makes it difficult to connect parts. Second, the number of people in each image is uncertain, who may come out in any location or proportion. OpenPose [1, 2] achieves the state-of-art performance, but there are still some common failure cases in practice including the rare, missing or incorrect location detection and overlap

© Springer Nature Singapore Pte Ltd. 2019
J. Chen et al. (Eds.): KSS 2019, CCIS 1103, pp. 18–27, 2019.
https://doi.org/10.1007/978-981-15-1209-4_2

of pose, part detections shared by two people, connecting error of two people, which usually results from incomplete person identification.

Instance segmentation provides a way to solve the person identification problem, which plays an important role in multi-person estimation. Instance segmentation can lead multi-person pose estimation model to focus on the important regions and eliminate noise effects. So the performance of instance segmentation directly affect the results of multi-person estimation. Now instance segmentation is a challenging task because it requires accurate detection of all objects in the image and accurate segmentation of each target instance. Mask R-CNN [3] represents the state-of-art position in terms of instance segmentation.

The goal of this study is to improve the pose estimation accuracy of multi-person. Therefore, we propose an improved two-stage multi-person pose estimation model to promote the person identification for further improving the performance of multi-person pose estimation. Mask R-CNN is used for person identification as the first stage of multi-person pose estimation model. The second stage, processed images or videos with identified people only are fed into OpenPose model for multi-person pose estimation. The comparative experiments show a significant improvement than original model. More people are captured so as to get more pose estimation, which directly affect the performance of behavior recognition.

In this study, Sect. 2 reviews the background and related work on pose estimation. The proposed improved two-stage multi-person pose estimation model is discussed in Sect. 3. Section 4 illustrates the experimental results and analysis. We conclude the study and suggest topics for future research in Sect. 5.

2   Related Work

2.1   Human Pose Estimation

The traditional single person pose estimation is based on local observation and spatial correlation. The spatial model of joint attitude is either based on the tree structure graph model [4] or non-tree model [5]. With the development of deep learning, especially Convolutional Neural Networks (CNNs), human pose estimation has been developed rapidly and significantly. Many pose estimation methods are used to estimate human posture in terms of legs, arms, hands and faces [6, 7]. However, all of the methods are processed on the assumption of a single person. For multi-person pose estimation, OpenPose is proposed by Carnegie Mellon University Team and achieves the state-of-art performance. Part Affinity Fields (PAFs) is used in OpenPose, which is a representation of a set of flow fields that encode unstructured pairwise relationships between a variable numbers of human body components [1]. The image output of pose estimation is shown in Fig. 1. The number from 0 to 25 in first image in Fig. 1 corresponds to nose, neck, right shoulder, right elbow, right wrist, left shoulder, left elbow, left wrist, mid hip, right hip, right knee, right ankle, left hip, left knee, left ankle, right eye, left eye, right ear, left ear, left big toe, left small toe, left heel, right big toe, right small toe, right heel and background. The second and third image stands for the pose information of hand and face. The applications of pose estimation is popular. Paschalis *et al.* [15] presented a method for the real-time estimation of the full 3D pose of multi-person hands based on OpenPose and achieved a state of the art. Research on multi-person pose estimation is hot recently. Miaopeng Li *et al.* [16] proposed a multi-person pose estimation using bounding box constraint and LSTM for single-image. They found that LSTM had superiority in handling input quality degradation in videos and stabilized the sequential outputs successfully.

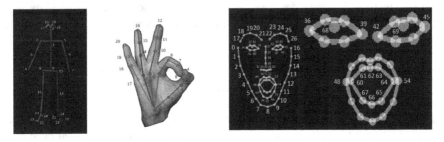

Fig. 1. Information of pose estimation

2.2 Instance Segmentation

The instance segmentation consists of two parts: The first is to classify and calculate the visual task of target detection, whose goal is to classify each target, and use the boundary box to locate it. The second is semantic segmentation, which aims to classify each pixel into a fixed pixel group of the category without distinguishing the target instance. Mask R-CNN achieves state-of-art performance on the instance segmentation. It belongs to the detection-based method, which gets the regions of each instance, and then predict the masks of each region. Some examples are shown in Fig. 2. It is developed as an extension of Faster R-CNN [8]. Mask R-CNN uses Fully Convolution Network (FCN) for semantic segmentation in each proposal box of Faster R-CNN, and the segmentation task is simultaneously performed with positioning and classification tasks. Also, Region of Interest (RoI) Align is introduced to replace RoI Pool in Faster R-CNN, which improve the mask accuracy significantly.

Fig. 2. Some examples of Mask R-CNN

3 Two-Stage Multi-person Pose Estimation Model

In this section, the improved two-stage multi-person pose estimation model is illustrated in detail. The framework of our proposed model is illustrated in Fig. 3.

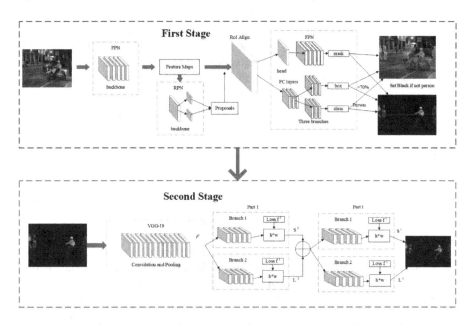

Fig. 3. Framework of two-stage multi-person pose estimation model

Our proposed model can be divided into two stages, which is introduced as follows:

(1) The first stage of our model realizes the recognition, segmentation and extraction of multi-person instances based on Mask R-CNN. On this basis, this method extracts all the people instances with background removed in image or video, and the number of people can be calculated.

   The first stage can be divided into three parts. The first is the backbone part, which is a series of convolutional layers for extracting feature maps of images. Feature Pyramid Network (FPN) [9] is used as backbone. It integrate feature maps from the bottom to the top, so as to make full use of the characters extracted in each stage. The second part is Region Proposal Network (RPN), the candidate area generation network. It is used to recommend the regions of interest to the network, which is also an important part of the faster-RCNN. The third is the network to classify candidate areas. Two rounding methods are used in RoI pooling, which may not cause a big error for classification and object detection, but have a significant influence on instance segmentation. It is obvious visually when the mask is misaligned. RoI Align uses bilinear interpolation instead for smaller error. After RoI Align in Mask R-CNN, there is a "head" section, whose main purpose is to enlarge the output dimension of RoI Align, so that it can be more accurate in predicting the mask. During the training session at mask branch, average binary cross-entropy loss is adopted instead of softmax loss of FCN and output mask for each class.

(2) The second stage of our model puts the focus on the pose estimation based on segmented image or video with people only. First, VGG-19 is used for feature extraction of input image from the first stage [10], outputting a feature map F. In part 1, two branches are utilized for prediction. The first branch is deployed for the confidence maps (S), which is also the classical method, CPM [10]. In addition, the branch of PAFs skeleton point direction (L) is added on the second branch. The output of two branches will be fused and transfer to the next part. The following part is similar to the above and last t phases, getting more accurate result. Finally, the output S and L of the network is obtained. The loss functions of S and L at t phase is listed as follows:

$$f_S^t = \sum_{j=1}^{J} \sum_{P} W(p) \cdot \left\| S_j^t(P) - S_j^*(p) \right\|_2^2 \tag{1}$$

$$f_L^t = \sum_{c=1}^{C} \sum_{P} W(p) \cdot \left\| L_c^t(P) - L_c^*(p) \right\|_2^2 \tag{2}$$

where S_j^* is the ground truth of confidence map and L_c^* is the ground truth of PAFs. W is the binary mask of $W(p) = 0$ when the annotation of image position P is missing. Masks are used to avoid punishing the real true positive in the training process. The supervision at each phase solves the problem of gradient disappearance by periodically supplementing gradient. The overall objective is

$$f = \sum_{t=1}^{T} (f_S^t + f_L^t) \tag{3}$$

After two stages, the extracted image with multi-person pose estimation can be merged into original image.

4 Experimental Result

In order to examine the performance of our proposed model, it is evaluated on two benchmarks for multi-person pose estimation: (1) Mall dataset [11] and (2) ShanghaiTech dataset [12]. These two datasets include images in diverse real-world scenarios such as crowding, scale variation, occlusion, and contact. The description of the datasets are shown in Table 1.

Chen et al. [11] collected Mall data set with different lighting conditions and crowd density. Data sets were collected using monitoring cameras of the shopping center. In addition to varying levels of density, it has different patterns of activity, stationary and moving crowds. In addition, the scenarios of the dataset have severe angular distortions,

Table 1. Description of experimental datasets.

Datasets	Image number	Image size	Min	Max	Average	Total count
Mall	2000	320 * 240	13	53	–	62,325
ShanghaiTech dataset part A	482	Varied	33	3139	501	241,677
ShanghaiTech dataset part B	716	768 * 1024	9	578	123	88,488

leading to great changes of persons and instances. The dataset provides the occlusion problems caused by scene objects such as houseplants along the walking path.

Zhang [12] introduced a new large-scale population statistics data set, consisting of 1,198 images and 330,165 annotation headers. It is the largest data set with large annotated location of people, and it consists of part A and part B. Part A includes 482 images randomly selected at the Internet, while part B is from the Shanghai metropolitan area on the street. The dataset successfully establish the challenging datasets with different scenario forms and different density levels. The complexity of the dataset provides a new view on person identification and a new opportunity for more applicability and extensible models.

The programs were run on a DELL Precision 7820 Tower workstation with a Nvidia GeForce RTX 2080 Ti (11 GB RAM) GPU, a dual Intel Xeon Silver CPU (12 cores, 2.1 GHz) 4116 and 8 * 8G DDR3 RAM.

The performance measures for multi-person identification are three indicators, which are mean square error (MSE), mean absolute error (MAE) and coefficient of determination (R^2). Suppose that y_i is the predicted value of the *ith* instances, and y_h is the corresponding true value, N is the number of instances, the three indicators are defined as follows:

$$MSE = \frac{1}{N}\sum_{i=1}^{N}(y_i - \widehat{y_h})^2 \tag{4}$$

$$MAE = \frac{1}{N}\sum_{i=1}^{N}|(y_i - \widehat{y_h})| \tag{5}$$

$$R^2 = 1 - \frac{\sum_{i=1}^{N}(y_i - \widehat{y_h})^2}{\sum_{i=1}^{N}(y_i - \widehat{y})^2} \tag{6}$$

Data preprocessing plays an important role in model evaluation. We remove images without person and extremely blurred. The model weights are trained through COCO dataset and tested on the three experimental datasets. The proposed improved two-stage multi-person pose estimation model is compared with original model OpenPose in Table 2. MSE, MAE and R^2 are computed based on the prediction values of models and the ground truth values of images. From the results we can find that our proposed model can effectively improve the prediction performance in terms of MSE, MAE, R^2, Mean values. Our proposed method reduces the MSE, MAE by around 27.38%, 21.57% and increases R^2, Mean values by 49.80% and 96.91% on average. The result of best model are highlighted with bold font. Also, we show the box plots in Fig. 4 for more intuitive comparisons, where we can find our proposed method, ITMPE, have better adaptability to multi-person scenarios.

Figures 5, 6 and 7 show randomly chosen images of three datasets, Mall dataset, ShanghaiTech dataset part A and ShanghaiTech dataset part B. In Figs. 5, 6 and 7, a_1, a_2 are the pose estimation of original model OpenPose and b_1, b_2 are the pose estimation of our proposed model. They are listed from the left to the right for the convenience of comparison. From the images, we can obviously find that more people are detected and less detection problems emerge with our model. The parts circled in red indicate that our model has superior performance in false identification, missing identification and so on.

Table 2. Performance comparison for person identification of two models for three dataset.

Dataset	Model	MSE	MAE	R^2	Mean	std.	Range
Mall	OpenPose	284.82	16.28	−4.91	14.876	**4.53**	4–31
	ITMPE	**184.15**	**11.95**	**−2.82**	**19.5**	6.57	**3–61**
Shanghai_part_A	OpenPose	531542.4	513.65	−1.05	29.39	**21.34**	0–121
	ITMPE	**470872.5**	**457.66**	**−0.81**	**86.50**	30.70	**10–127**
Shanghai_part_B	OpenPose	15169.69	86.075	−0.72	43.98	**10.66**	19–78
	ITMPE	**9801.29**	**62.65**	**−0.11**	**72.73**	19.68	**26–127**

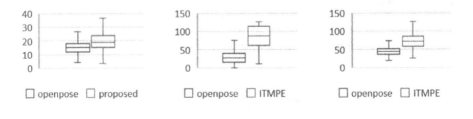

a. Mall dataset   b. Shanghai_part_A dataset   c. Shanghai_part_B dataset

Fig. 4. Box plots of two models for datasets

From Table 2, we can find that there is no obvious difference in the person identification on the Mall dataset because of the small number of people in image, but our proposed model, ITMPE, performs better in terms of overlap of pose and misidentification, which results from noise elimination at the first stage. For the part A and part B of ShanghaiTech dataset, ITMPE have overwhelming superiority than OpenPose because of its advantages in human identification. Some examples are shown in Figs. 6 and 7.

Fig. 5. Comparison of OpenPose and our proposed model for Mall dataset (Color figure online)

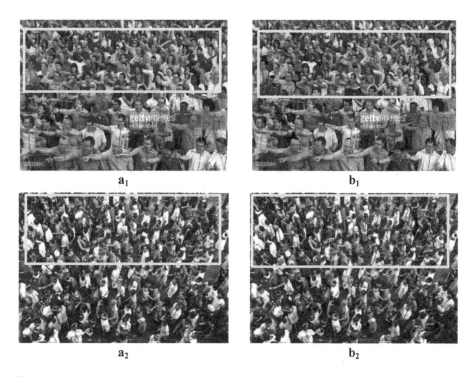

Fig. 6. Comparison of OpenPose and our proposed model for ShanghaiTech dataset part A (Color figure online)

$$\mathbf{a_1} \qquad\qquad\qquad \mathbf{b_1}$$

$$\mathbf{a_2} \qquad\qquad\qquad \mathbf{b_2}$$

Fig. 7. Comparison of OpenPose and our proposed model for ShanghaiTech dataset part B (Color figure online)

5 Conclusion

Previously, there exists underestimate or misidentification conditions in multi-person pose estimation, especially complex scenarios. They are usually caused by the rare, missing or incorrect location detection and overlap of pose and so on. In our study, we propose an improved two-stage multi-person pose estimation model to improve the performance of multi-person pose estimation. The experimental results show the efficiency to estimate pose of multi-person. Our proposed model reduces the MSE, MAE by around 27.38%, 21.57% and increases R^2, Mean values by 49.80% and 96.91% on average. Moreover, more people are captured and given the pose estimation. This will directly affect the performance of behavior recognition. In the future, we will add experimental data in different scenarios and compare the proposed model with other latest models. In addition, we will further study behavior recognition based on the multi-person pose estimation.

References

1. Cao, Z., Simon, T., Wei, S.E., Sheikh, Y.: Realtime multi-person 2D pose estimation using part affinity fields. In: Proceedings of 30th IEEE Conference on Computer Vision and Pattern Recognition, CVPR 2017 (2017)

2. Zhu, X., Jiang, Y., Luo, Z.: Multi-person pose estimation for PoseTrack with enhanced part affinity fields. In: ICCVW (2017)
3. He, K., Gkioxari, G., Dollar, P., Girshick, R.: Mask R-CNN. In: Proceedings of IEEE *International Conference* on *Computational* Vision (2017)
4. Yang, Y., Ramanan, D.: Articulated human detection with flexible mixtures of parts. TPAMI **35**, 2878–2890 (2013)
5. Dantone, M., Gall, J., Leistner, C., Van Gool, L.: Human pose estimation using body parts dependent joint regressors. In: CVPR (2013)
6. Chu, X., Yang, W., Ouyang, W., Ma, C., Yuille, A.L., Wang, X.: Multi-context attention for human pose estimation. In: CVPR (2017)
7. Tang, W., Yu, P., Wu, Y.: Deeply learned compositional models for human pose estimation. In: ECCV (2018)
8. Girshick, R.: Fast R-CNN. In: Proceedings of IEEE International Conference on Computer Vision (2015). https://doi.org/10.1109/iccv.2015.169
9. Lin, T.Y., Dollár, P., Girshick, R., He, K., Hariharan, B., Belongie, S.: Feature pyramid networks for object detection. In: Proceedings of 30th IEEE Conference on Computer Vision and Pattern Recognition, CVPR 2017 (2017)
10. Simonyan, K., Zisserman, A.: VGG-16. arXiv Preprint (2014)
11. Wei, S.-E., Ramakrishna, V., Kanade, T., Sheikh, Y.: Convolutional pose machines. In: CVPR (2016)
12. Chen, K., Loy, C.C., Gong, S., Xiang, T.: Feature Mining for Localised Crowd Counting (2012)
13. Idrees, H., Saleemi, I., Seibert, C., Shah, M.: Multi-source multi-scale counting in extremely dense crowd images. In: Proceedings of the IEEE Computer Society Conference on Computer Vision and Pattern Recognition (2013)
14. Zhang, Y., Zhou, D., Chen, S., Gao, S., Ma, Y.: Single-image crowd counting via multi-column convolutional neural network. In: Proceedings of the IEEE Computer Society Conference on Computer Vision and Pattern Recognition (2016)
15. Panteleris, P., Oikonomidis, I., Argyros, A.: Using a single RGB frame for real time 3D hand pose estimation in the wild. In: Proceedings of 2018 IEEE Winter Conference on Applications of Computer Vision, WACV 2018 (2018)
16. Li, M., Zhou, Z., Liu, X.: Multi-person pose estimation using bounding box constraint and LSTM. IEEE Trans. Multimedia (2019)

A Fuzzy AHP-TOPSIS Approach for Selecting the Multimodal Freight Transportation Routes

Kwanjira Kaewfak[1,2](✉) [ID], Van-Nam Huynh[1] [ID], Veeris Ammarapala[2] [ID], and Chayakrit Charoensiriwath[3] [ID]

[1] School of Knowledge Science, Japan Advanced Institute of Science and Technology, Ishikawa 923-121, Japan
{kwanjira,huynh}@jaist.ac.jp
[2] School of Management Technology, Sirindhorn International Institute of Technology, Thammasat University, Pathumthani 12000, Thailand
[3] NECTEC, National Science and Technology Development Agency, Pathumthani 12120, Thailand

Abstract. Multimodal transportation route selection strategy has become an important component in the main logistics and transportation. Route selection relies upon decision-based on real industry data and expert judgments. This paper proposes Fuzzy Analytic Hierarchy Process (AHP) and Fuzzy Technique for Order of Preference by Similarity to Ideal Solution (TOPSIS) for prioritizing effectively the multimodal transportation routes to improve logistics system performance by constructing the possible routes considering transport cost, time, risk, and quality factors. Fuzzy AHP is used to determine weights for evaluation criteria and Fuzzy TOPSIS is used to aid the ranking of possible route alternatives. The empirical case study of coal manufacturing is conducted to illustrate a proposed methodology that enables to provide a more accurate, practical, and systematic decision support tool.

Keywords: Multimodal freight transportation · Route selection · Fuzzy set theory · Fuzzy AHP · Fuzzy TOPSIS

1 Introduction

Freight transportation is a main supply chain system and logistics component to ensure the efficient movement and timely availability of raw materials and finished products [1,2]. Multimodal Transportation has been considered as a key component of modern logistics systems, especially for long distance transportation and large volume. Multimodal transportation, as defined by Multimodal Transport Handbook published by UNCTAD, is the transport of products by

Supported by Supported by SIIT-JAIST-NECTEC Dual Doctoral Degree Program Scholarship.

© Springer Nature Singapore Pte Ltd. 2019
J. Chen et al. (Eds.): KSS 2019, CCIS 1103, pp. 28–46, 2019.
https://doi.org/10.1007/978-981-15-1209-4_3

different transportation modes from one point to a destination point where one of the carriers organizes the whole transport. Route selection is a fundamental problem of logistics management, which makes great contributions to real life applications [3,4]. Most of studies that focused on research fields, such as air logistics, railway networks, shipping logistics, and multimodal transportation networks have emphasized it [4–10]. Most previous studies pertaining to route selection problems adopt mathematical models such as stochastic programming or integer programming in order to maximize service quality [4,11–16]. However, these researches rarely concern the criteria that cannot be expressed as real data, such as fuel prices, transport capacity, vehicle traffic, operational risk, flexibility etc. This paper bridges this gap by utilizing fuzzy AHP and fuzzy TOPSIS approaches, which handles both qualitative and quantitative criteria in order to evaluate the reasoning behind route selection, as well as to define the optimal freight route alternatives. The coal manufacturing located on central area of Thailand will be used as a case study analysis.

The rest of this paper is organized as follows. In next section, a literature review is conducted on Multiple Criteria Decision Making (MCDM) model including fuzzy AHP and fuzzy TOPSIS. The proposed technique of Fuzzy AHP and Fuzzy TOPSIS are described in Sect. 3. Section 4 presents case illustration. Section 5 covers the conclusion, limitation and further study.

2 Literature Review

Multiple Criteria Decision Making (MCDM) methods are approaches to structure information and decision evaluation in formal problems with multiple, conflicting goals [17,18]. In the literature, many authors have applied MCDM methods to evaluate and select the transportation routes including the Analytic Hierarchy Process (AHP) [19]. Ammarapala et al. [20] utilized AHP method select potential rural roads to support cross-border shipment. In their study, Kengpol et al. [21] applies the AHP to determine the weights of the criteria and linguistic terms of assessment grades for route selection between Thailand and Vietnam multimodal freight routes. The AHP has been widely applied in several studies to used technique for prioritizing the multi-criteria decision system. It has unique advantage when important of element are difficult to compare.

However, AHP method provides a structured framework for setting priorities on each level of the hierarchy using pairwise comparisons that are quantified using 1–9 scales [19]. The traditional AHP requires exact judgments from experts. Since the uncertainty and vagueness of the experts' opinion is the prominent characteristic of the problem, AHP is difficult to assign an exact numerical value in pairwise comparison as the prioritization process is complex and subjective. Thus, instead of using an exact numerical number, FAHP approach utilizes Triangular Fuzzy Number (TFN) to express the pairwise comparison of decision elements [21]. This is absolutely what lead researchers to propose a fuzzy set theory integrating with AHP method, which has been used to numerous of uncertainty situation [17,22–24].

TOPSIS is a well-known technique for classic MCDM as well proposed by Hwang and Yoon [25]. The underlying logic of TOPSIS is to identify a solution from a finite set of alternatives that the chosen alternative should have the shortest distance from the positive ideal solution and the farthest distance from the negative ideal solution. Many researchers used it to solve the fuzzy MCDM problem [17,26–29]. Because attribute weighting in TOPSIS has high subjectivity and decision makers can give weight for attributes directly, without regarding the consistency of the weight value [30,31]. Moreover, Fuzzy AHP and TOPSIS FAHP and TOPSIS methods can be used together for complex decision problems. Therefore, this research proposed the fuzzy AHP integrating with fuzzy TOPSIS. Fuzzy AHP is applied to determine the preference weights and then used fuzzy TOPSIS to meet the overall objective and rank the alternatives routes [17].

3   The Proposed Technique

For problem formulation, the following notation is adopted from general of MCDM theory. Let us consider MCDM problem, where $A_i(i = 1, 2, .., m)$ are possible alternatives multimodal routes and $C_j(j = 1, 2, .., n)$ are set of criteria (cost, time, risk, and quality) Finally, it is assumed that a decision maker is able to reflect the importance of the criteria using a set of n weights $w_1, w_2, ..., w_n$. X_{ij} is performance rating of alternative A_i with respect to criteria C_j. A multiple criteria (or multiple attribute) decision making (MCDM) problem can be represented using a decision matrix as the one shown in following Table 1.

Table 1. The decision matrix of MCDM problem

	c_1	c_2	$\cdots$	c_n
	w_1	w_2	$\cdots$	w_n
A_1	x_{11}	x_{12}	$\cdots$	x_{1n}
A_2	x_{21}	x_{22}	$\cdots$	x_{2n}
$\vdots$	$\vdots$	$\vdots$	$\ddots$	$\vdots$
A_m	x_{m1}	x_{m2}	$\cdots$	x_{mn}

This research tries to evaluate the logistics system performance by constructing the possible routes considering transport cost, time, risk, and quality factors. After reviewing the literature, we set the criteria that building the hierarchy structure. Based on the evaluation criteria, this research lists the eight possible multimodal routes. The research employs fuzzy AHP to determine the weight for main and evaluation criteria, while fuzzy TOPSIS is adapted to rank the options.

We use fuzzy AHP to fuzzify hierarchical analysis by allowing fuzzy numbers for pairwise comparison and find the fuzzy preference weight [26]. In this section, we briefly review the concepts of fuzzy hierarchical evaluation. Then, we will introduce the computational process of fuzzy AHP in detail.

3.1 Fuzzy AHP and Extent Analysis

Establishing Fuzzy Number. We adopt the TFNs in applications due to their computational simplicity, and useful fuzzy information processing. The definitions and algebraic operations are described as follows. TFNs can be defined by a triplet (l, m, u) and its membership function $\mu_A(x)$ can be defined by Eq. (1) [32]. From Eq. (1) l and u mean the lower and upper bounds of the fuzzy number A, and m is the modal value for A.

$$
\mu_A(x) = \begin{cases}
\dfrac{x-l}{m-l}, & l \leq x \leq m \\
\dfrac{u-x}{u-m}, & m \leq x \leq u \\
0, & otherwise
\end{cases} \tag{1}
$$

(where x is a real number of the interval $[0,1]$ and l, m, u are real numbers.)
Define two TFNs A and B by the triplets $A = (l_1, m_1, u_1)$ and $B = (l_2, m_2, u_2)$
Then: *Addition* $: A \oplus B = (l_1, m_1, u_1)(+)(l_2, m_2, u_2) = (l_1+l_2, m_1+m_2, u_1+u_2)$
Multiplication $: A \otimes B = (l_1, m_1, u_1)(l_2, m_2, u_2) = (l_1 l_2, m_1 m_2, u_1 u_2)$,
Inverse $= (l_1, m_1, u_1)^{-1} \approx (\frac{1}{u_1}, \frac{1}{m_1}, \frac{1}{l_1})$

Determining the Linguistic Variables. We translate the linguistic terms used by decision makers to express the comparative judgments among the main criteria with the respect to overall goal, and sub-criteria with respect to main criteria into TFNs. The arrangement of comparison matrices will be as below:

$$
\tilde{A} - (\tilde{a}_{ij})_{n*n} = \begin{bmatrix}
(1,1,1) & (l_{12}, m_{12}, u_{12}) & \dots & (l_{1n}, m_{1n}, u_{1n}) \\
(l_{21}, m_{21}, u_{21}) & (1,1,1) & \dots & (l_{2n}, m_{2n}, u_{2n}) \\
\vdots & \vdots & \ddots & \vdots \\
(l_{n1}, m_{n1}, u_{n1}) & (l_{n2}, m_{n2}, u_{n2}) & \dots & (1,1,1)
\end{bmatrix} \tag{2}
$$

The range of values used in fuzzy AHP utilized the below scale shown in Table 2 [26, 33]. We define by three parameters of the symmetric triangular fuzzy number, the left point, middle point, and right point of the range over which the function is defined.

Table 2. Scale of relative importance used in the pairwise comparison matrix

Uncertainty judgement	Triangular fuzzy scale	Triangular fuzzy reciprocal scale
Equally important	$(1,1,1)$	$(1,1,1)$
Weakly important	$(2,3,4)$	$(1/4, 1/3, 1/2)$
Fairly important	$(4,5,6)$	$(1/6, 1/5, 1/4)$
Strongly important	$(6,7,8)$	$(1/8, 1/7, 1/6)$
Absolutely important	$(8,9,10)$	$(1/10, 1/9, 1/8)$

Aggregation of the Preferences. For aggregation the preference of t decision makers in order to construct the final pairwise comparison matrix, we proposed the weight aggregation technique based on geometric mean of preferences [26,34], the following equations are described:

$$LW_{ij} = ((\prod_{t=1}^{T} LW_{ijt})^1/t$$

$$MW_{ij} = ((\prod_{t=1}^{T} MW_{ijt})^1/t \qquad (3)$$

$$UW_{ij} = ((\prod_{t=1}^{T} UW_{ijt})^1/t$$

Prioritizing the Key Transportation Factors. Various methods have been developed to handle fuzzy comparison matrices. Among the methods, Chang [32]'s the extent analysis method has been adopted in various number of applications order to computational simplicity [3]. Extend analysis is the technique used to prioritize the weights. In this paper, we proposed the extend analysis method by Chang [32], which derives crisp weights for fuzzy comparison matrices. The algorithm of this method can be described as follows:

Let $\tilde{A} = (\tilde{a}_{ij})_{nm}$ be a fuzzy pairwise comparison matrix, where $(\tilde{a}_{ij}) = (l_{ij}, m_{ij}, u_{ij})$. According to method of Chang [30] extend analysis, each object is taken and extend analysis for each goal, g_i is performed respectively. It means that it is possible to obtain the values of m extent analyses that can be demonstrated as $M_{gi}^1, M_{gi}^2, \ldots, M_{gi}^m, i = 1, 2, \ldots, n$ where all the M_{gi}^j (j = 1, 2, .., m) are TFNs.

The step of Chang's extent analysis can be given as in the following:

Step 1. The value of fuzzy synthetic extent with the respect to the ith object is defined as:

$$S_i = \sum_{j=1}^{m} M_{gi}^1 \otimes \left[\sum_{i=1}^{n} \sum_{j=1}^{m} M_{gi}^1 \right]^{-1} \qquad (4)$$

To obtain $\sum_{j=1}^{m} M_{gi}^1$ performed the fuzzy addition operation of m extend analysis values for a particular matrix such that

$$\sum_{j=1}^{m} M_{gi}^1 = [\sum_{j=1}^{m} l_j, \sum_{j=1}^{m} m_j, \sum_{j=1}^{m} u_j] \qquad (5)$$

and to obtain $\sum_{i=1}^{n} \sum_{j=1}^{m} M_{gi}^1$, performed the fuzzy addition operation of $M_{gi}^1(j = 1, 2, \ldots, m)$ values such that

$$\sum_{i=1}^{n} \sum_{j=1}^{m} M_{gi}^1 = [\sum_{j=1}^{n} l_i, \sum_{j=1}^{n} m_i, \sum_{j=1}^{n} u_i] \qquad (6)$$

The inverse of the vector in Eq. (16) can be computed as:

$$\left[\sum_{i=1}^{n}\sum_{j=1}^{m}M_{gi}^{1}\right]^{-1} = \left[\frac{1}{\sum_{j=1}^{n}u_i}, \frac{1}{\sum_{j=1}^{n}m_i}, \frac{1}{\sum_{j=1}^{n}l_i}\right] \tag{7}$$

Step 2. The degree of possibility of $M_2 = (l_2, m_2, u_2)$ and $M_1 = (l_1, m_1, u_1)$ is defined as:

$$V(M_2 \geq M_1) = \sup_{y \geq x}\lfloor min(\mu_{M1}(x), \mu_{M2}(y))\rfloor \tag{8}$$

and can be equivalently expressed as follows:

$$V(M_2 \geq M_1) = hgt(M_1 \cap M_2) = \mu_{M2}(d) = \begin{cases} 1 & m_2 \geq m_1 \\ 0 & l_1 \geq u_2 \\ \frac{l_1 - u_2}{(m_2 - u_2) - (m_1 - l_1)} & otherwise \end{cases} \tag{9}$$

where d is the ordinate of the highest intersection point D between μ_{M1} and μ_{M2} (See in Fig. 1.)

To compare M_1 and M_2, we need both the values of $V(M_2 \geq M_1)$ and $V(M_1 \geq M_2)$.

Step 3. The degree possibility for a convex fuzzy number to be greater than k convex fuzzy numbers M_i $(i = 1, 2, ..., k)$ can be defined as,

$$\begin{aligned} V(M &\geq M_1, M_2, ..., M_k) \\ &= V[(M_{\geq}M_1) \ and \ (M \geq M_2] \ and \ ... \ and \ (M \geq M_k)] \\ &= min V(M \geq M_i), i = 1, 2, 3, .., k \end{aligned} \tag{10}$$

Assume that,

$$d'(A_i) = min V(S \geq S_k) \tag{11}$$

For $k = 1, 2, ..., n$ and & k. The weight vector can be given by the following formula:

$$W' = (d'(A_1), d'(A_2), ..., d'(A_n))^T \tag{12}$$

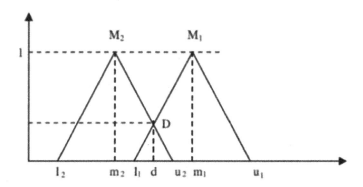

Fig. 1. The interaction between M_1 and M_2.

Step 4. Normalization step, the normalized weight vectors are given as:

$$W' = (d'(A_1), d'(A_2), ..., d'(A_n))^T, \tag{13}$$

where W is a non-fuzzy number

After determining the fuzzy attribute weights, we will be used them as the weight of qualitative criteria for the fuzzy TOPSIS methodology.

3.2   Integration with Fuzzy TOPSIS for Ranking the Alternatives

Based on the Hwang and Yoon [25] research, fuzzy TOPSIS is a method to identify a solution form a finite point. The chosen points are the shortest point in distance from positive ideal and the farthest point in distance from negative ideal solution. The aim of this research focus on the fuzzy TOPSIS method as the continuation of fuzzy AHP method. In the real world problem, it is difficult that decision makers can provide the precise numerical judgments about evaluating in each criteria. Therefore, fuzzy TOPSIS are employed in this research. The decision makers need to come a conclusion for their decision using importance level of criteria rating. Then, the aggregated opinion of decision makers is considered to be equally important and consider as a MCDM problem.

As it is stated, there are $m\&$ possible alternatives called $A = \{A_1, A_2, ..., A_m\}$ which are evaluated by four decision makers based on four main criteria, $C = \{C_1, C_2, ..., C_n\}$. The performance ratings of each expert D_k $(k = 1, 2, .., K)$ for each alternative A_i $(i = 1, 2, .., m)$ with respect to criteria C_j $(j = 1, 2, .., n)$ are denoted by $\tilde{R}_k = \tilde{X}_{ijk}$ $(i = 1, 2, .., m; j = 1, 2, .., n; k = 1, 2, .., K)$ membership function $\mu\tilde{R}k(x)$. The scale used for alternative rating is given in Table 3.

Calculating the Aggregate Fuzzy Rating for the Alternatives. If the rating of all experts are explained as TFNs $\tilde{R} = (ab, bk, ck)$, $k = 1, 2, ..K$ then the aggregated rating is given by $\tilde{R} = (a, b, c)$ $k = 1, 2, .., K$ where

$$a = \min_k \{a_k\}, b = \frac{1}{k} \sum_{k=1}^{K} bk, c = \max_k \{c_k\} \tag{14}$$

Table 3. Fuzzy linguistic variable for alternative rating [28].

Linguistic variables	Corresponding TFNs
Very poor	$(1, 1, 3)$
Poor	$(1, 3, 5)$
Fair	$(3, 5, 7)$
Good	$(5, 7, 9)$
Very good	$(7, 9, 11)$

But If the rating of all experts are explained as TFNs $\tilde{X}_{Ijk} = (a_{ijk}, b_{ijk}, c_{ijk})$, $i = 1, 2, ..m$, $j = 1, 2, ..n$ then the aggregated rating is given by $\tilde{X}_{Ij} = (a_{ij}, b_{ij}, c_{ij})$ where

$$a_{ij} = \min_k \{a_{ij}\}, b = \frac{1}{k} \sum_{k=1}^{K} b_{ijk}, c = \max_k \{c_{ijk}\} \tag{15}$$

Constructing the Normalize Fuzzy Decision Matrix. After building the fuzzy decision matrix for alternatives The raw data will be normalized using linear scale transformation to bring the various criteria scales into a comparative scale. The normalized fuzzy decision matrix $\tilde{R}$ is given by: $R = [r_{ij}]_{mn}, i = 1, 2, ..m; j = 1, 2, .., n$ where

$$\tilde{r}_{ij} = (\frac{a_{ij}}{c_j}, \frac{b_{ij}}{c_j}, \frac{c_{ij}}{c_j}) \text{ and } c_j^* = max_i c_{ij} \text{ (benefit criteria)} \tag{16}$$

$$\tilde{r}_{ij} = (\frac{a_j^-}{c_{ij}}, \frac{a_j^-}{b_{ij}}, \frac{a_j^-}{a_{ij}}) \text{ and } a_j^- = min_i a_{ij} \text{ (cost criteria)} \tag{17}$$

We compute the weighted normalized matrix $\tilde{v}$ for criteria by multiplying the weight (w_j) of evaluation criteria with the normalized fuzzy decision matrix $\tilde{r}_{ij}$.

$$\tilde{V} = [\tilde{v}_{ij}]_{mn}, i = 1, 2, ..m; j = 1, 2, .., n \text{ where } \tilde{v}_{ij} = \tilde{r}_{ij}(.)W_j \tag{18}$$

Note that $\tilde{v}_{ij}$ is a TFN represented by $(\tilde{a}_{ijk}, \tilde{b}_{ijk}, \tilde{c}_{ijk})$.

Determining the Fuzzy Ideal Solution (FPIS) and Fuzzy Negative Ideal Solution (FNIS). We calculate the FPIS and FNIS of the alternatives by:

$$A^* = (\tilde{v}_1^*, \tilde{v}_2^*, ..., \tilde{v}_n^*) \text{ where } \tilde{v}_j^* = (\tilde{c}_j^*, \tilde{c}_j^*, \tilde{c}_j^*) \text{ and } \tilde{c}_j^* = max_i\{\tilde{c}_{ij}\} \tag{19}$$

$$A^* = (\tilde{v}_1^-, \tilde{v}_2^-, ..., \tilde{v}_n^-) \text{ where } \tilde{v}_j^* = (\tilde{a}_j^-, \tilde{a}_j^-, \tilde{a}_j^-) \text{ and } \tilde{a}_j^- = min_i\{\tilde{a}_{ij}\} \tag{20}$$

After that calculate the distance of each alternative from FPIS and FNIS. The distance (d_i^+, d_i^-) of each weighted alternative $i = 1, 2, .., m$ is computed as follows:

$$d_i^+ = \sum_{j=1}^{n} dv(\tilde{v}_{ij}, \tilde{v}^*_j), i = 1, 2, ..., m \tag{21}$$

$$d_i^- = \sum_{j=1}^{n} dv(\tilde{v}_{ij}, \tilde{v}^-_j), i = 1, 2, .., m \tag{22}$$

Calculating the Closeness Coefficient (CCi) of Each Alternatives and Ranking the Alternatives. The closeness coefficient represents the distance to the fuzzy positive ideal solution (A^*) and the fuzzy negative ideal solution (A^-) simultaneously. The closeness coefficient of each alternative is computed as:

$$CC_i = \frac{d_i^-}{d_i^- + d_i^+} \tag{23}$$

The last step, the different alternatives are ranked according to CC_i in descending order.

4   Case Illustration

4.1   Descriptive Information

The data taken into account for the modeling include the current possible coal logistic multimodal routes in Thailand. The research utilizes the data that collected from decision makers in order to identify areas of study and appropriate multimodal transportation routes. There are 8 possible transportation routes that are combinations of several different modes of transport (e.g. rail, sea and road). These 8 possible multimodal transportation routes can be as shown in Table 4.

Table 4. The possible transportation routes

No	Alternative routes
A1	Koh si-chang - Laem Chabang = ... - Kaeng Khoi Cement Plant Saraburi Province
A2	Koh si-chang + Laem Chabang = ... - Kaeng Khoi Cement Plant Saraburi Province
A3	Koh si-chang + Pasak River + ... - Kaeng Khoi Cement Plant Saraburi Province
A4	Koh si-chang + Pasak River + ... - Kaeng Khoi Cement Plant Saraburi Province
A5	Koh si-chang + Pasak River + ... - Kaeng Khoi Cement Plant Saraburi Province
A6	Koh si-chang + Bang Pa Kong - ..... - Kaeng Khoi Cement Plant Saraburi Province
A7	Koh si-chang + Bang Pa Kong - ..... - Kaeng Khoi Cement Plant Saraburi Province
A8	Koh si-chang + Bang Pa Kong - ..... - Kaeng Khoi Cement Plant Saraburi Province

Note: + is ship transport, = is train transport, - is truck transport *

Four main criteria i.e. transport cost, transport time, transport risk and quality and more twenty-eight sub-criteria are identified through literature review and intensive discussion with decision group member. The model is constructed by combining two methods of MCDM: fuzzy AHP and fuzzy TOPSIS.

The four expert panel comprising of logistics and transportation specialist are formed to evaluate the optimal multimodal route. Total eight possible multimodal routes. The FAHP method is used first in order to allow for determination of weight vectors for each transport factors individually. Then the hierarchy structure is formed such the goal or objective is the first level, main criteria in the second level, sub-criteria in the third level and the possible routes are the alternative options. All factors that identified by literature review and expert opinion are described in Table 5., where the categories of main factors are marked as factors $= \{C, T, R, Q\}$ and sub-criteria are denoted as $= \{C1, C2, C3, .., Q7\}$ respectively.

4.2 Fuzzy AHP for Route Selection Criteria Determination

In this phase the decision group is asked to make pairwise comparisons of four main criteria and twenty-eight sub-criteria by using linguistic variables as shown in Table 1. As it is given in Tables 6 and 7, the geometric mean of preferences value are computed to obtain the pairwise comparison matrix of each sub-criteria (Due to space limitation, the pairwise comparison matrix of transport cost and time are only given here.).

Table 5. Route selection factors.

Main criteria	Criteria code	Sub criteria
Cost (C)	C1	Mode of transport [33]
	C2	Total distances [35]
	C3	Fuel prices [36]
	C4	Amount of cargo [37]
	C5	Transport capacity [37]
	C6	Type of goods and size of shipment [37]
	C7	Missed transshipment [37]
Time (T)	T1	Vehicle traffic
	T3	Vehicle speed
	T4	Period of time
	T5	Geography [35]
	T6	Storage time during transshipment
	T7	Vehicle departure/ arrival rate
	T2	Disrupting transport system in general [37]
Risk (R)	R1	Freight damaged risk [38,39]
	R2	Infrastructure and equipment risk [38–40]
	R3	Operational risk [38–41]
	R4	Security risk [38–42]
	R5	Environment risk [38–40]
	R6	Policy and Law risk [38–40]
	R7	Financial risk [41]
Quality (Q)	Q1	Trip safety
	Q2	Flexibility
	Q3	On-time delivery performance
	Q4	Product quality [42]
	Q5	Easy booking and contract
	Q6	Frequency of transportation
	Q7	Multimodal service

Table 6. Pairwise comparison matrix of the transport cost.

	C1	C2	C3	C4	C5	C6	C7
C1	(1.000, 1.000, 1.000)	(0.250, 0.333, 0.500)	(2.000, 3.000, 4.000)	(3.000, 4.000, 5.000)	(1.000, 2.000, 3.000)	(1.000, 2.000, 3.000)	(1.000, 1.000, 1.000)
C2	(2.000, 3.000, 4.000)	(1.000, 1.000, 1.000)	(1.000, 1.000, 1.000)	(1.000, 2.000, 3.000)	(1.000, 2.000, 3.000)	(1.000, 2.000, 3.000)	(2.000, 3.000, 4.000)
C3	(0.250, 0.333, 0.500)	(1.000, 1.000, 1.000)	(1.000, 1.000, 1.000)	(2.000, 3.000, 4.000)	(2.000, 3.000, 4.000)	(2.000, 3.000, 4.000)	(3.000, 4.000, 5.000)
C4	(0.250, 0.333, 0.500)	(0.333, 0.500, 1.000)	(0.250, 0.333, 0.500)	(1.000, 1.000, 1.000)	(1.000, 1.000, 1.000)	(1.000, 1.000, 1.000)	(0.167, 0.200, 0.250)
C5	(0.333, 0.500, 1.000)	(0.333, 0.500, 1.000)	(0.250, 0.333, 0.500)	(1.000, 1.000, 1.000)	(1.000, 1.000, 1.000)	(1.000, 1.000, 1.000)	(0.167, 0.200, 0.250)
C6	(0.333, 0.500, 1.000)	(0.333, 0.500, 1.000)	(0.250, 0.333, 0.500)	(1.000, 1.000, 1.000)	(1.000, 1.000, 1.000)	(1.000, 1.000, 1.000)	(0.200, 0.250, 0.333)
C7	(1.000, 1.000, 1.000)	(0.250, 0.333, 0.500)	(0.250, 0.333, 0.500)	(4.000, 5.000, 6.000)	(4.000, 5.000, 6.000)	(3.000, 4.000, 5.000)	(1.000, 1.000, 1.000)

Table 7. Pairwise comparison matrix of the transport time.

	T1	T2	T3	T4	T5	T6	T7
T1	(1.000, 1.000, 1.000)	(6.000, 7.000, 0.800)	(1.000, 1.000, 1.000)	(2.000, 3.000, 4.000)	(1.000, 2.000, 3.000)	(1.000, 2.000, 3.000)	(1.000, 1.000, 1.000)
T2	(0.125, 0.143, 0.167)	(1.000, 1.000, 1.000)	(0.167, 0.200, 0.250)	(0.250, 0.333, 0.500)	(0.250, 0.333, 0.500)	(0.333, 0.500, 1.000)	(0.333, 0.500, 1.000)
T3	(1.000, 1.000, 1.000)	(4.000, 5.000, 6.000)	(1.000, 1.000, 1.000)	(2.000, 3.000, 4.000)	(4.000, 5.000, 6.000)	(4.000, 5.000, 6.000)	(1.000, 1.000, 1.000)
T4	(0.250, 0.333, 0.500)	(2.000, 3.000, 4.000)	(0.250, 0.333, 0.500)	(1.000, 1.000, 1.000)	(0.250, 0.333, 0.500)	(1.000, 1.000, 1.000)	(1.000, 1.000, 1.000)
T5	(0.333, 0.500, 1.000)	(2.000, 3.000, 4.000)	(0.167, 0.200, 0.250)	(2.000, 3.000, 4.000)	(1.000, 1.000, 1.000)	(1.000, 1.000, 1.000)	(1.000, 1.000, 1.000)
T6	(0.333, 0.500, 1.000)	(1.000, 2.000, 3.00)	(0.167, 0.200, 0.250)	(1.000, 1.000, 1.000)	(1.000, 1.000, 1.000)	(1.000, 1.000, 1.000)	(1.000, 1.000, 1.000)
T7	(1.000, 1.000, 1.000)	(1.000, 2.000, 3.000)	(1.000, 1.000, 1.000)	(1.000, 1.000, 1.000)	(1.000, 1.000, 1.000)	(1.000, 1.000, 1.000)	(1.000, 1.000, 1.000)

The values of fuzzy synthetic extents with respect to the criteria weights are determined by Eqs. (4)–(6). These synthesis extents are fuzzy criteria weights of route selection factors (See in Tables 8, 9, 10 and 11). The final weight results obtain from the calculations based on previous equation are presented in Fig. 2.

As it is shown in Tables 8, 9, 10 and 11, triangular fuzzy numbers are determined and they will be used as the weight of qualitative criteria for fuzzy TOPSIS methodology.

Table 8. Fuzzy criteria weights of cost

Decision criteria	Weights
C1	$(0.105, 0.186, 0.314)$
C2	$(0.121, 0.230, 0.393)$
C3	$(0.128, 0.209, 0.341)$
C4	$(0.043, 0.063, 0.105)$
C5	$(0.046, 0.070, 0.123)$
C6	$(0.047, 0.072, 0.128)$
C7	$(0.112, 0.169, 0.268)$

Table 9. Fuzzy criteria weights of time

Decision criteria	Weights
T1	$(0.148, 0.230, 0.330)$
T2	$(0.029, 0.044, 0.077)$
T3	$(0.208, 0.284, 0.386)$
T4	$(0.063, 0.089, 0.133)$
T5	$(0.084, 0.120, 0.179)$
T6	$(0.069, 0.097, 0.141)$
T7	$(0.104, 0.135, 0.172)$

4.3 Evaluation of Routes Selection and Determines Final Rank by Fuzzy TOPSIS

The expert panel members are asked to construct a fuzzy evaluation matrix by using linguistic variables presented in Table 3. It is established by comparing alternative routes under each of factors separately. Then convert linguistic terms into corresponding TFNs and construct the fuzzy evaluation matrix. Aggregated fuzzy weight of alternatives are calculated using Eq. (15) and presented in

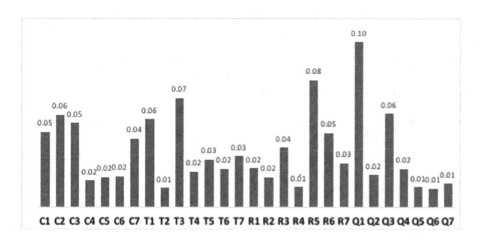

Fig. 2. The Final weight of transportation factors

Table 10. Fuzzy criteria weights of risk

Decision criteria	Weights
R1	$(0.061, 0.096, 0.156)$
R2	$(0.054, 0.077, 0.109)$
R3	$(0.104, 0.155, 0.218)$
R4	$(0.030, 0.048, 0.081)$
R5	$(0.221, 0.333, 0.471)$
R6	$(0.135, 0.195, 0.266)$
R7	$(0.077, 0.113, 0.162)$

Table 11. Fuzzy criteria weights of quality

Decision criteria	Weights
Q1	$(0.297, 0.431, 0.605)$
Q2	$(0.050, 0.080, 0.129)$
Q3	$(0.181, 0.244, 0.328)$
Q4	$(0.059, 0.095, 0.155)$
Q5	$(0.031, 0.048, 0.081)$
Q6	$(0.031, 0.047, 0.070)$
Q7	$(0.038, 0.060, 0.091)$

Table 12. In this research, cost, time and risk factors are termed as cost criteria and normalization performed by Eq. (17). On the contrary, quality is indexed as benefit criteria and using Eq. (16) to normalize the fuzzy decision matrix (See Table 13.). Next step is to obtain a fuzzy weighted evaluation matrix by using the weights calculated by fuzzy AHP.(See Tables 8, 9, 10 and 11.). Thus, the weighted normalized are established using Eq. (18) which is presented in Table 14.

In this research, the fuzzy ideal positive solution (FPIS) and fuzzy ideal negative solution (FNIS) can be determined as follows respectively by using Eqs. (21)–(22). $\tilde{v}^* = (0, 0, 0)$ and $\tilde{v}^- = (1, 1, 1)$. The calculation of the distance of each alternative route from FPIS and FNIS, and the final results are summarized in Table 15.

4.4   Result and Discussions

To select suitable multimodal routes, the hybrid fuzzy AHP-TOPSIS approach are utilized in this research that ranking process are made it more comprehensive and systematic. This hybrid approach is intended to select the optimal route effectively. This will be achieved by implementing and considering the related transport factors. Total 28 factors and 8 alternative routes choice are identified by literature review and expert opinion. The weight of each criteria calculated by fuzzy AHP and by using these criteria weight the alternative routes ranked by fuzzy TOPSIS method. The fuzzy TOPSIS results of the route selection under study are shown in Table 15. The evaluation of route selection is realized and according to CC_i value ranking are A5-A4-A7-A3-A8-A1-A2-A6 from most suitable to least. Hence, Thai case manufacturing should consider and implement into their logistics process.

Table 12. The aggregated fuzzy decision matrix for route selection.

	C1	C2	C3	.....	Q5	Q6	Q7
A1	(5.00, 8.50, 11.00)	(5.00, 7.50, 11.00)	(1.00, 2.50, 7.00)	.....	(3.00, 6.00, 9.00)	(7.00, 9.00, 11.00)	(5.00, 7.00, 9.00)
A2	(5.00, 8.50, 11.00)	(5.00, 7.50, 11.00)	(1.00, 2.50, 7.00)	.....	(3.00, 6.00, 9.00)	(7.00, 9.00, 11.00)	(5.00, 7.00, 9.00)
A3	(3.00, 7.50, 11.00)	(3.00, 6.00, 9.00)	(1.00, 2.00, 5.00)	.....	(3.00, 6.00, 9.00)	(5.00, 7.50, 11.00)	(5.00, 7.00, 9.00)
A4	(1.00, 6.00, 9.00)	(3.00, 6.50, 11.00)	(1.00, 6.00, 9.00)	.....	(1.00, 2.50, 7.00)	(1.00, 2.50, 9.00)	(1.00, 3.00, 7.00)
A5	(1.00, 6.00, 9.00)	(3.00, 6.50, 11.00)	(1.00, 6.00, 9.00)	.....	(1.00, 2.50, 7.00)	(1.00, 2.50, 9.00)	(1.00, 3.00, 7.00)
A6	(3.00, 7.00, 11.00)	(3.00, 5.50, 9.00)	(1.00, 3.00, 7.00)	.....	(3.00, 5.00, 7.00)	(3.00, 5.50, 9.00)	(3.00, 6.00, 9.00)
A7	(5.00, 8.50, 11.00)	(3.00, 5.00, 7.00)	(1.00, 3.50, 7.00)	.....	(3.00, 5.00, 7.00)	(3.00, 5.50, 9.00)	(3.00, 6.00, 9.00)
A8	(5.00, 8.00, 11.00)	(3.00, 5.00, 7.00)	(1.00, 3.50, 7.00)	.....	(3.00, 5.00, 7.00)	(3.00, 5.50, 9.00)	(3.00, 6.00, 9.00)

Table 13. Normalized fuzzy decision matrix for route selection.

	C1	C2	C3	.....	Q5	Q6	Q7
A1	(0.09, 0.12, 0.20)	(0.273, 0.40, 0.60)	(0.14, 0.40, 1.00)	.....	(0.33, 0.67, 1.00)	(0.64, 0.82, 1.00)	(0.56, 0.78, 1.00)
A2	(0.09, 0.12, 0.20)	(0.273, 0.40, 0.60)	(0.14, 0.40, 1.00)	.....	(0.33, 0.67, 1.00)	(0.64, 0.82, 1.00)	(0.56, 0.78, 1.00)
A3	(0.09, 0.12, 0.33)	(0.33, 0.50, 1.00)	(0.20, 0.50, 1.00)	.....	(0.33, 0.67, 1.00)	(0.64, 0.82, 1.00)	(0.56, 0.78, 1.00)
A4	(0.11, 0.17, 1.00)	(0.27, 0.46, 1.00)	(0.11, 0.17, 1.00)	.....	(0.11, 0.28, 0.78)	(0.09, 0.50, 0.82)	(0.11, 0.33, 0.78)
A5	(0.11, 0.17, 1.00)	(0.27, 0.46, 1.00)	(0.11, 0.17, 1.00)	.....	(0.11, 0.28, 0.78)	(0.09, 0.23, 0.82)	(0.11, 0.33, 0.78)
A6	(0.09, 0.14, 0.33)	(0.33, 0.54, 1.00)	(0.14, 0.33, 1.00)	.....	(0.33, 0.56, 0.78)	(0.27, 0.50, 0.82)	(0.33, 0.67, 1.00)
A7	(0.09, 0.12, 0.20)	(0.43, 0.60, 1.00)	(0.14, 0.30, 1.00)	.....	(0.33, 0.56, 0.78)	(0.27, 0.50, 0.82)	(0.33, 0.67, 1.00)
A8	(0.09, 0.13, 0.20)	(0.43, 0.60, 1.00)	(0.14, 0.30, 1.00)	.....	(0.33, 0.56, 0.78)	(0.27, 0.50, 0.82)	(0.33, 0.67, 1.00)

Table 14. Weighted normalized fuzzy decision matrix for route selection

	C1	C2	C3	.....	Q5	Q6	Q7
A1	(0.01, 0.02, 0.06)	(0.03, 0.09, 0.24)	(0.02, 0.08, 0.34)	.....	(0.01, 0.03, 0.08)	(0.02, 0.04, 0.07)	(0.02, 0.05, 0.09)
A2	(0.01, 0.02, 0.06)	(0.03, 0.09, 0.24)	(0.02, 0.08, 0.34)	.....	(0.01, 0.03, 0.08)	(0.02, 0.04, 0.07)	(0.02, 0.05, 0.09)
A3	(0.01, 0.02, 0.10)	(0.04, 0.12, 0.39)	(0.03, 0.10, 0.34)	.....	(0.01, 0.03, 0.08)	(0.02, 0.04, 0.07)	(0.02, 0.05, 0.09)
A4	(0.01, 0.02, 0.10)	(0.04, 0.12, 0.39)	(0.03, 0.10, 0.34)	.....	(0.00, 0.01, 0.06)	(0.00, 0.02, 0.06)	(0.00, 0.020.07)
A5	(0.01, 0.03, 0.31)	(0.03, 0.11, 0.39)	(0.01, 0.03, 0.34)	.....	(0.00, 0.01, 0.06)	(0.00, 0.01, 0.06)	(0.00, 0.020.07)
A6	(0.01, 0.03, 0.10)	(0.04, 0.13, 0.39)	(0.02, 0.07, 0.34)	.....	(0.01, 0.03, 0.06)	(0.01, 0.02, 0.06)	(0.01, 0.040.09)
A7	(0.01, 0.02, 0.06)	(0.05, 0.14, 0.39)	(0.02, 0.06, 0.34)	.....	(0.01, 0.03, 0.06)	(0.01, 0.02, 0.06)	(0.01, 0.040.09)
A8	(0.01, 0.02, 0.06)	(0.05, 0.14, 0.39)	(0.02, 0.06, 0.34)	.....	(0.01, 0.03, 0.06)	(0.01, 0.02, 0.06)	(0.01, 0.040.09)

Table 15. Closeness coefficient (CC_i) and ranking of alternative routes.

Code	Route	d_i^+	d_i^-	(CC_i)	Rank
A1	Koh si-chang - Laem Chabang = ... - Kaeng Khoi Cement Plant Saraburi Province	25.73	2.89	0.101	6
A2	Koh si-chang + Laem Chabang = ... - Kaeng Khoi Cement Plant Saraburi Province	25.78	2.88	0.100	7
A3	Koh si-chang + Pasak River + ... - Kaeng Khoi Cement Plant Saraburi Province	25.67	3.02	0.105	4
A4	Koh si-chang + Pasak River + ... - Kaeng Khoi Cement Plant Saraburi Province	25.72	3.17	0.109	2
A5	Koh si-chang + Pasak River + ... - Kaeng Khoi Cement Plant Saraburi Province	25.64	3.35	0.115	1
A6	Koh si-chang + Bang Pa Kong - ..... - Kaeng Khoi Cement Plant Saraburi Province	25.82	2.87	0.099	8
A7	Koh si-chang + Bang Pa Kong - ..... - Kaeng Khoi Cement Plant Saraburi Province	25.71	3.07	0.106	3
A8	Koh si-chang + Bang Pa Kong - ..... - Kaeng Khoi Cement Plant Saraburi Province	25.73	2.99	0.104	5

5   Conclusion, Limitation and Further Study

Multimodal transportation is considered as an important development in making local industry and international trade more efficient and competitive. However, route selection is a main problem in multimodal transportation process. Thus the need arises to overcome this problem by providing the solution. It is difficult to implement all solutions at the same time due to various constraints, therefore ranking the alternative route is essential to step-wise implementation of these solution. This research are presenting a scientific framework to rank the alternative multimodal routes by using a hybrid multiple criteria decision technique which combines fuzzy AHP and fuzzy TOPSIS. Human are often uncertain in assigning the evaluation scores and the uncertainty and vagueness of the experts' opinion is the prominent characteristic of the problem. Hence, AHP and TOP-SIS are performed in fuzzy environment. Fuzzy AHP is used to get the weights of multimodal transport factors that consider transportation cost, time, quality and risk while fuzzy TOPSIS is integrate to rank the alternative routes. The weights obtain from fuzzy AHP are included in fuzzy TOPSIS computations and the alternative routes priorities are determined. The empirical case study is presented to demonstrate the applicability of the proposed framework. The case study of coal manufacturing is conducted to illustrate a proposed method-ology. The coal Industry is one of the important industries that is suitable for multimodal transportation because it is a nonperishable product, with long lead time of transportation and being transported in large volume. The case study of the domestic freight route in multimodal transportation originating from Koh Sri Chang, Chonburi Province in Thailand to the destination of a Cement fac-tory in Saraburi Province, Thailand. The data are collected from interview and brainstorming of experts in order to identify areas of study and appropriate multimodal transportation routes. Four experts from different areas of works associated with multimodal transportation of coal industries were interviewed. Through the literature review the expert opinion total 28 factors and 8 possible multimodal routes are identified and hybrid fuzzy AHP-TOPSIS framework are performed to rank the alternative routes. The result shows route 5 (Koh si-chang + Pasak River + Nakornluang Port - Route 3008 - Route 3470 - Route 3041 - Route 362 - Mittraphap Road - Kaeng Khoi Cement Plant Saraburi Province)* towards transport factors adoption is the highest rank route. As the results shown in the empirical case study, it is that the proposed method is practical for ranking and finding the optimal multimodal freight transportation route.

However, there are a number of limitations in this empirical case study. Route selection cannot be determined solely on the total weight. These factors can be adjusted. Apart from that, the interviewees are majorly from the experts in logis-tics and transportation area in Thailand that might bring various perspectives regarding factors affecting route selection. The majority of data acquired in this research is also based on Thailand economic and environment. Therefore, the factors must be adjusted before applying in other countries.

These solution ranking helps organization to decide their optimal multimodal freight transportation route effectively. This proposed technique gives a new valid and reliable approach for prioritizing the alternative route under related factor consideration. It is the main contribution of this research. For further research, the results can be compared with the other fuzzy MCDM techniques such as fuzzy ELECTRE, fuzzy PROMETHEE etc. Moreover, this research plans to develop a new algorithm to solve the transport problem at a large scale.

References

1. Crainic, T.G.: Handbook of Transportation Science. Kluwer Academic Publishers, Norwell (2003)
2. SteadieSeifi, M., Dellaert, N.P., Nuijten, W., Van Woensel, T., Raoufi, R.: Multimodal freight transportation planning: a literature review. Eur. J. Oper. Res. **233**, 1–15 (2014)
3. Park, Y.I., Lu, W., Nam, T.H., Yeo, G.T.: Terminal vitalization strategy through optimal route selection adopting CFPR methodology. Asian J. Shipp. Logist. **35**, 41–48 (2019)
4. Huynh, N., Fotuhi, F.: A new planning model to support logistics service providers in selecting mode, route, and terminal location. Pol. Marit. Res. **20**, 67–73 (2013)
5. Banomyong, R., Beresford, A.: Multimodal transport: the case of Laotian garment ex-porters. Int. J. Phys. Distrib. Logist. Manag. **31**(9), 663–685 (2001)
6. Krile, S.: Efficient heuristic for non-linear transportation problem on the route with multiple ports. Pol. Marit. Res. **20**(4), 80–86 (2013)
7. Balakrishnan, A., Karsten, C.V.: Container shipping service selection and cargo routing with transshipment limits. Eur. J. Oper. Res. **263**(2), 652–663 (2017)
8. Raza, Z.: The commercial potential for LNG shipping between Europe and Asia via the Northern Sea Route. J. Marit. Res. **11**(2), 67–79 (2014)
9. Sheffi, Y., Mahmassani, H., Powell, W.B.: A transportation network evacuation model. Transp. Res. Part A Gen. **16**(3), 209–218 (1982)
10. Qu, L., Chen, Y.: A hybrid MCDM method for route selection of multimodal transportation network. In: Sun, F., Zhang, J., Tan, Y., Cao, J., Yu, W. (eds.) ISNN 2008. LNCS, vol. 5263, pp. 374–383. Springer, Heidelberg (2008). https://doi.org/10.1007/978-3-540-87732-5_42
11. Carbone, V., Martino, M.D.: The changing role of ports in supply-chain management: an empirical analysis. Marit. Policy Manag. **30**(4), 305–320 (2003)
12. Cabral, A.M.R., Ramos, F.S.: Cluster analysis of the competitiveness of container ports in Brazil. Transp. Res. Part A Policy Pract. **69**, 423–431 (2014)
13. Dang, V.L., Yeo, G.T.: A competitive strategic position analysis of major container ports in Southeast Asia. Asian J. Shipp. Logist. **33**(1), 19–25 (2017)
14. Feng, L., Notteboom, T.: Small and medium-sized ports (SMPs) in multi-port gateway regions: the role of Yingkou in the logistics system of the Bohai sea. In: Notteboom, T. (ed.) Current Issues in Shipping, Ports and Logistics, pp. 543–563. University Press Antwerp, Brussels (2011)
15. Feng, L., Notteboom, T.: Peripheral challenge by small and medium sized ports (SMPs) in multi-port gateway regions: the case study of northeast of China. Pol. Marit. Res. **20**, 55–66 (2013)
16. Vujić, M., Skorput, P., Mandžuka, B.: Multimodal route planners in maritime environment. Pomorstvo **29**(1), 1–7 (2015)
17. Rostamzadeh, R., Sofian, S.: Prioritizing effective 7Ms to improve production systems performance using fuzzy AHP and fuzzy TOPSIS (case study). Expert Syst. Appl. **38**, 5166–5177 (2011)

18. Rahman, M.A., et al.: Selection of the best inland waterway structure: a multicriteria decision analysis approach. Water Resour. Manag. **29**, 2733–2749 (2015)
19. Saaty, T.L.: The Analytic Hierarchy Process: Planning, Priority Setting, Resources Allocation. McGraw, New York (1980)
20. Ammarapala, V., Chinda, T., Pongsayaporn, P., Ratanachot, W., Punthutaecha, K., Janmonta, K.: Cross-border shipment route selection utilizing analytic hierarchy process (AHP) method, p. 7 (2018)
21. Angelo, P.M., Furuichi, T., Ishii, N.: A fuzzy analytic network process for multicriteria evaluation of contaminated site remedial countermeasures. J. Environ. Manag. **88**, 479–495 (2008)
22. Bottani, E., Rizzi, A.: A fuzzy multi-attribute framework for supplier selection in an e-procurement environment. Int. J. Logist. Res. Appl. **8**(3), 249–266 (2005)
23. Chan, F.T.S., Kumar, N., Tiwari, M.K., Lau, H.C.W., Choy, K.L.: Global supplier selection: a fuzzy-AHP approach. Int. J. Prod. Res. **46**(14), 3825–3857 (2008)
24. Mikhailov, L.: Fuzzy analytical approach to partnership selection in formation of virtual enterprises. Omega **30**(5), 393–401 (2002)
25. Hwang, C.L., Yoon, K.: Multiple Attribute Decision Making: Methods and Applications. Springer, Heidelberg (1981). https://doi.org/10.1007/978-3-642-48318-9
26. Sun, C.-C.: A performance evaluation model by integrating fuzzy AHP and fuzzy TOPSIS methods. Expert Syst. Appl. **37**, 7745–7754 (2010)
27. Mandic, K., Delibasic, B., Knezevic, S., Benkovic, S.: Analysis of the financial parameters of Serbian banks through the application of the fuzzy AHP and TOPSIS methods. Econ. Model. **43**, 30–37 (2014)
28. Patil, S.K., Kant, R.: A fuzzy AHP-TOPSIS framework for ranking the solutions of Knowledge Management adoption in Supply Chain to overcome its barriers. Expert. Syst. Appl. **41**, 679–693 (2014)
29. Taylan, O., Bafail, A.O., Abdulaal, R.M.S., Kabli, M.R.: Construction projects selection and risk assessment by fuzzy AHP and fuzzy TOPSIS methodologies. Appl. Soft Comput. **17**, 105–116 (2014)
30. Zhang, Z., Guo, C.: Deriving priority weights from intuitionistic multiplicative preference relations under group decision-making settings. J. Oper. Res. Soc. **68**, 1582–1599 (2018)
31. Zhang, Z., Kou, X., Yu, W., Guo, C.: On priority weights and consistency for incomplete hesitant fuzzy preference relations. Knowl. Based Syst. **143**, 115–126 (2017)
32. Chang, D.: Applications of the extent analysis method on fuzzy AHP. Eur. J. Oper. Res. **95**, 649–655 (1996)
33. Gumus, A.-T.: Evaluation of hazardous waste transportation firms by using a two step fuzzy-AHP and TOPSIS methodology. Expert. Syst. Appl. **36**(2), 4067–4074 (2009)
34. Jaiswal, R., Ghosh, N.C., Lohani, A., Thomas, T.: Fuzzy AHP based multi crteria decision support for watershed prioritization. Water Resour. Manag. **29**, 4205–4227 (2015)
35. Rodrigue, J.P., Comtois, C., Slack, B.: The Geography of Transport Systems. Routledge, New York (2008)
36. Novák, P., Popesko, B.: Cost variability and cost behaviour in manufacturing enterprises. Econ. Sociol. **7**(4), 89–103 (2014)
37. Andersson, M., Berglund, M., Flodén, J., Persson, C., Waidringer, J.: A method for measuring and valuing transport time variability in logistics and cost benefit analysis. Res. Transp. Econ. **66**, 59–69 (2017)

38. Kengpol, A., Tuammee, S.: The development of a decision support framework for a qualitative risk assessment in multimodal green logistics: an empirical study. Int. J. Prod. Res. **54**, 1020–1038 (2016)
39. Kengpol, A., Tuammee, S., Tuominen, M.: The development of a framework for route selection in multimodal transportation. Int. J. Logist. Manag. **25**(3), 581–610 (2014)
40. Kiba-Janiak, M.: Opportunities and threats for city logistics development from a local authority perspective. J. Econ. Manag. **28**(2), 23–39 (2017)
41. Ibrahimovic, S., Franke, U.: A probabilistic approach to IT risk management in the Basel regulatory framework: a case study. J. Financ. Regul. Compliance **25**(2), 176–195 (2017)
42. Trond, S.N., Fallah, Z.: Risk perceptions, fatalism and driver behaviors in Turkey and Iran. Saf. Sci. **59**, 187–192 (2013)

Evolution Analysis of Newcomer-Task Network Structure of Enterprise Information System: A Case Study of a Shipbuilding Enterprise

Yanhong Peng[1], Jun Yin[1,2(✉)], Shi-lun Ge[1], and Peng Liu[1]

[1] School of Economics and Management,
Jiangsu University of Science and Technology, Zhenjiang 212003, China
bamhill@163.com
[2] Service Manufacturing Model and Information Research Center,
Jiangsu University of Science and Technology, Zhenjiang 212003, China

Abstract. With the implementation of the enterprise information system, newcomers of information system often feel that their work efficiency decreases in the process of using unfamiliar system functions, which affects the utilization rate of information system as well as their performance. As the core groups of information system implementation, understanding the behavior of newcomers is the starting point of deeply grasping the operation mechanism of information system user relationship network in enterprises, thus it is of great value to study the behavior of newcomers in information system. According to the user system usage data in a ship enterprise information system, the newcomer-task network structure of the information system is analyzed with the complex network methods. It is found that the evolution of the newcomer-task network has better restored the implementation cycle of the enterprise information system, that is, it has undergone a new period, a growing period and a stable period in the system. The relationship between newcomers and other users has gradually dispersed from the beginning to form a small group and finally gathered.

Keywords: Enterprise information system · Newcomers · Network structure evolution

1 Introduction

Newcomers in enterprises information system are a group that needs special attention. On the one hand, they will experience a steep learning curve when trying to understand the system, and often struggle with the most basic system functions, this process often causes them a lot of complaints, resulting in unnecessary work delays and additional workload [1]; On the other hand, the implementation of the system will bring them changes in work tasks and business processes. When they perceive these changes as

The National Natural Science Foundation of China (7187010393, 71331003), Jiangsu Postgraduate Research Innovation Project (SJKY19_2602).

© Springer Nature Singapore Pte Ltd. 2019
J. Chen et al. (Eds.): KSS 2019, CCIS 1103, pp. 47–63, 2019.
https://doi.org/10.1007/978-981-15-1209-4_4

negative or potential threats, they will be more likely to take actions that hinder the implementation of information systems [2]. Under this background, it is very important to analyze the newcomer behavior of the enterprise information system and correctly manage the newcomers of the information system to improve the utilization rate of the system.

Newcomers are greatly affected by the environment. Google conducted an experiment in which they promptly encouraged a group of managers with new employees by sharing suggestions on how to help new employees adapt to the new environment. Under the experimental conditions, managers' initiative in helping new employees adapt to the new environment increased by 14%, and the registration speed of new employees increased by 10% [3]. Previous studies have found that newcomers tend to rely more on colleagues, leaders and other help in the face of changes in work tasks and business processes [4]. Kowtha found that new employees can easily adapt to their roles under the guidance of senior colleagues, and the social support of ordinary colleagues and supervisors can promote the transformation of their roles [5]; Elis et al. proposed that managers' perception of new employees' initiative is related to their friendly managerial behavior, and managers' evaluation and response to new employees' initiative will support the adjustment of new employees [6]. Thus, it may be more effective to analyze newcomers from the perspective of social network structure; Moreover, different stages of enterprise information system will present different stages characteristics, and each stage has a clear task [7], the arrival of newcomers at different stages faces different problems. Therefore, it is necessary to consider the stage characteristics of newcomers' environment when analyzing them.

This paper uses construct the social network of newcomers of information system using the log data of the information system user operating the same functional module to. Furthermore, the dynamic evolution of the network (micro and macro) is used to analyze the characteristics and changing rules of newcomers' usage behavior in the whole information system cycle.

2   Literature Review

With the introduction of enterprise information systems, the enterprise environment has become decentralized and virtualized, newcomers of information systems are increasingly emphasizing the use of systems to connect with other users in the organization to obtain information [8], thereby improving the quality of use and work efficiency. Table 1 summarizes the research in other areas from the perspective of newcomers, including enterprises, schools and hospitals. Combined with the relevant researches, we find that newcomers often need to go through a process of social adaptation due to the lack of effective guidance and the support of experienced users when they come into contact with new things [6], which may cause difficulties in solving problems and making decisions, resulting in a decline in work efficiency [9]. Roland et al. [10] mentioned that new librarians can have a positive or negative impact on their work at different stages in difficult situations, and Sushil [11] suggested that both positive support and negative attack from managers in enterprises will affect the performance of newcomers.

Current research regards the process of newcomers' socialization as a social network problem. In the newcomers' social network, users are regarded as nodes, and they establish an edge by using the same function to connect with other users. Since the node set and edge set of newcomer task network change with time, it is called dynamic social network, which can be regarded as a static social network at each time point. Most of the existing research methods for newcomers are aimed at static networks or snapshots at a certain time, Lee believes that newcomers of technology specialty rely on direct link with users in structural holes to overcome the responsibility of new things and lack of resources, and that influential users and users in structural holes can force and restrain newcomers [12]; Chiu proposed that the interest, self-confidence and social support of freshmen have different effects on their academic persistence [13].

Table 1. Relevant researches in other areas from the perspective of newcomers

Perspective	Field	Authors	Objects	Factors	Conclusions
Newcomers	School	Roland et al. (2016) [10]	New Librarians	User attribute & Direct effect	Organizational support can have positive and negative effects on new librarians at different stages
		Chiu (2007) [13]	Freshman	User attribute & Direct effect	Interest in curriculum, self-confidence and social support has different effects on freshmen's academic persistence at different stages
	Hospital	Hesketh et al. (2003) [14]	Freshmen	Direct effect	Team communication, collaboration and personal development affect the educational process of new doctors
		McGuire et al. (1977) [15]	New patients	User attribute	New patients are more sensitive to environmental change; Staffs are more resistant to environmental change
	Enterprise	Delobbe et al. (2016) [16]	New employee	Direct effect	The behavior of new employees is influenced by supervisors and colleagues
		Smith et al. (2012) [17]	New employee	Direct effect	Positive and effective feedback from colleagues, as well as respect and fair treatment in the workplace can help new employees to invest
		Chang & Lin (2011) [18]	New Users	Direct effect & Indirect effect	Intranet system is helpful to newcomers information seeking and socialized behavior

From the above research, we can see that scholars have studied the influence of newcomers' attributes and network environment on their behavior from different fields, and most of them are based on static networks or snapshots at a certain time. In fact, a static network can't describe the process of growth, stability or decline of network members when they reach the final network location. Changes in network structure are still invisible and unexplored in the static network, but the dynamic evolution of social network can clearly observe the changes at every moment. According to the changes, the analysis results can be combined with the actual situation in the enterprise to formulate effective measures. Therefore, based on the existing research, it is very meaningful to analyze the dynamic evolution of the newcomers' environment.

3   Research Project Design

This paper designs an analytical framework for the evolution of newcomer-task network structure, which shows the research ideas and implementation steps more clearly, and uses network structure indicators to analyze the overall evolution of newcomer-task network from both micro and macro aspects. The framework for the evolution of newcomer-task network structure is shown in Fig. 1. It can be divided into the following five steps:

Step 1: Data collection. This paper chooses the log data as the data source in the information system of a shipbuilding enterprise and filters the data needed in this study by processing the abnormal data.

Step 2: A newcomer-task network build. The user-task network is a social network structure based on the user's use of system functions recently proposed by scholars [19]. Considering the information exchange between users using the same module, users are regarded as nodes and the relationship between users using the same function is regarded as the edge. This paper uses python software to build a newcomer-task network with newcomers (system experience less than 1 year) and other users using the same module.

Step 3: The overall evolution of the newcomer-task network. The overall evolution of the network is measured by the common indicators in the newcomer-task network. The micro-level analyzes the number of nodes and edges. At the macro level, the topology characteristics of the network structure, including the average aggregation coefficient, the average path length, the module degree and the degree distribution, are used to analyze the evolution of the newcomer-task network.

Step 4: Conclusions. According to the newcomer-task network evolution analysis results, combined with the current stage, it gives inspiration and suggestions for the development of enterprise information system usage behavior.

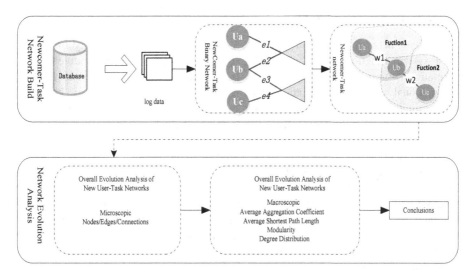

Fig. 1. Research Framework

4   Data Acquisition and Analysis Methods

4.1   Data Acquisition

Considering that a ship enterprise information system has been running since its implementation in 2011, it is a relatively successful information system. In the system implementation process, the continuous updating of the function modules is well adapted to the enterprise business, and the operation mode may be worth learning from other manufacturing enterprises or small and medium-sized enterprises. Therefore, this paper selects the system usage log data of the enterprise from 2011.11 to 2016.10 for 60 months. After processing the vacancy data, noise data, inconsistent data, duplicate data and incomplete data in the original data table, a total of 1,369,223 log data are shown in Fig. 2. 560 valid users, including 433 males, accounting for 77.3%, and 127 females, accounting for 22.7%, mainly distributed in the management department, the current department, and the basic production department, the auxiliary production department and the information implementation department.

ID	LoginName	LoginTime	ModuleName
308	User A	2011-11	Subject management of procurement contracts
309	User B	2011-11	Basic property maintenance
2337	User C	2011-12	Engineering code maintenance

Fig. 2. Information system usage log part information table

4.2 Network Construction Method

The user-task network is a recently proposed social network structure based on user's use of system functions by scholars [19]. The basis is the function module of the information system, and the research object is users who use the same module. Firstly, the functional units of information systems represent the corresponding business modules in the real world. Using the same functional units means that users may engage in similar work contents in enterprises and have similar or interdependent work responsibilities, so it is more likely to have direct or indirect potential connections [20]. Secondly, users who can use the same module in the system often have similar backgrounds, experiences, work environments and other common points. The information will be more efficient to transfer between users who have common points [21]. Therefore, the user-task network is not only an objective and real informal interpersonal network, but also an effective way for users to transfer knowledge. Its proposal can reflect the relationship between users of information systems to a certain extent. It can be used as a supplement to the traditional social network structure and has certain practical significance.

To sum up, the core of the user-task network is user and function. The foundation of constructing user-task network is to get user and corresponding functional modules, and form a two-mode network, which is expressed as three tuples in turn: $G = <U, F, e>$, U represents the user node, F is the functional node, and e is the functional link. After constructing the basic two -mode network, we need to determine the network node of the user task network, that is, using the function of the employees in the two-mode network, and taking the user node as the object, the two-mode network is projected into a user-task network with the user as the single vertex. To ensure the validity of user-task network exploration in different implementation stages, we add weights to the edges between network vertices, the user-task network is further represented as a binary group: $G = <N, E>$, N represents nodes, which represent each employee in the network and correspond to the actual employee. E represents the edge of the node connected to the node, including attributes such as weights.

Based on the principle of user-task network, this paper uses Python software to construct a newcomer-task network of information system monthly, in which the newcomer is a user with less than one year's system experience, and uses Gephi software to visualize the newcomer-task network and get the network indicators for each stage.

5 Network Evolution Analysis

From the perspective of the network, the evolution of the network can be divided into two levels, one is the micro level, and the other is the macro [22]. The micro level refers to the changes in network nodes and edges, and the changes in microstructure ultimately lead to changes in macro features. Firstly, the changes of nodes (users) and edges (the relationships between users using the same functions) in the newcomer-task network are analyzed in detail. Then the changes in network structure at the macro level of newcomer-task network are analyzed. To express conveniently, if there is no special proof, the node will be used to represent the users who use the information system, while edges expressing the relationship between users who use the same function in the information system.

5.1 Analysis of Network Evolution at Micro Level

Enterprise information system is a relatively complex system, in which there are various evolutionary phenomena, that is, users may enter the system at any time and continue to use, or quit the system to stop using, and may use it intermittently. As the nodes change, the edges change accordingly. Among them, employees may use the same module in the system because of similar work content, or they may use the same module to establish the connection because of different work needs and inconsistent work content. Therefore, we analyze the number of nodes and edges as well as newcomers' access to the system connection.

(1) Number of nodes/edges

The evolution of the number of nodes and edges per month of newcomers in the process of using the system is shown in Fig. 3. Among them, the change of newcomers' nodes has experienced a rapid growth, stabilization and decline process, that is, with the initial use of enterprise information systems to mature use, and the updating of business modules, the final it is in a relatively stable state. Enterprise information system is fully used by employees, and business module reaches a more popular state. At the same time, from the change in the number of employees using the same function, as the number of nodes changes, the situation of the connected edges basically changes similarly, that is, the entry of newcomers led to the smooth operation of the entire information system, the continuous use is to complete their work tasks (Fig. 4).

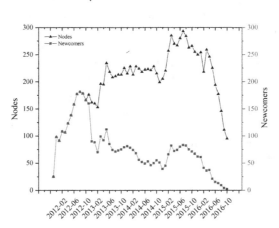

Fig. 3. Evolution diagram of number of newcomer-task nodes and number of newcomers

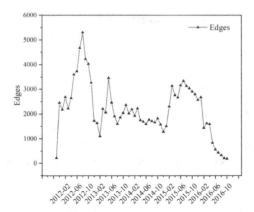

Fig. 4. Edge evolution diagram of newcomer-task network

(2) Connections

At the micro-level, this paper considers the addition and deletion of newcomer-task network nodes and the updating of edges. In the process of the changes, users' behavior of using the same function changes accordingly. On this basis, we further analyze the change of preferential connection behavior of newcomers in the process of using the information system, that is, the newly added nodes tend to connect with higher value nodes. For a newcomer, in addition to establishing contacts with acquaintances and colleagues, strangers tend to pay more attention to senior users, such as those who use more functional modules; there users have been concerned by a large number of users and have a higher degree of the node. Table 2 shows the connection between newcomers and all users every month, including the connection between newcomers and newcomers, newcomers and old users. Old users are defined as users with system experience greater than 1 year. Since a newcomer is defined as a user with less than one year of system experience, there is only a connection between the newcomer and the newcomer in the newcomer-task network in the previous year. In the next month, 2012.11, the existing newcomer part began to become an old user. At this stage, newcomers started to have new relations with old users. Similarly, we found that while newcomers entering the system are gradually decreasing and old users are increasing, most newcomers entering the information system rely more on establishing contact with experienced old users, but there are still connections with newcomers.

Table 2. Micro level analysis of newcomer-task networks

Time	All Users: Newcomers	New_New:New_Old	Time	All Users: Newcomers	New_New:New_Old
2011-11	25: 25	460 (100%): 0 (0%)	2014-05	218: 52	461 (19%): 1987 (81%)
2011-12	98: 98	6371 (100%): 0 (0%)	2014-06	222: 49	410 (19%): 1765 (81%)
2012-01	91:91	5950 (100%): 0 (0%)	2014-07	223: 53	480 (20%): 1931 (80%)
2012-02	108:108	5838 (100%): 0 (0%)	2014-08	221: 46	458 (19%): 1910 (81%)
2012-03	106: 106	5287 (100%): 0 (0%)	2014-09	228: 50	432 (18%): 1996 (82%)
2012-04	123: 123	5117 (100%): 0 (0%)	2014-10	215:55	512 (20%): 2033 (80%)
2012-05	138: 138	8091 (100%): 0 (0%)	2014-11	199:51	517 (22%): 1875 (78%)
2012-06	158: 158	7621 (100%): 0 (0%)	2014-12	205: 39	274 (16%): 1436 (84%)
2012-07	177: 177	8522 (100%): 0 (0%)	2015-01	220: 45	365 (16%): 1855 (84%)
2012-08	181: 181	9399 (100%): 0 (0%)	2015-02	257: 66	900 (29%): 2168 (71%)
2012-09	178: 178	7835 (100%): 0 (0%)	2015-03	285: 82	1227 (30%): 2821 (70%)
2012-10	166: 166	7226 (100%): 0 (0%)	2015-04	270: 72	1005 (27%): 2731 (73%)
2012-11	176: 160	5711 (92%): 477 (8%)	2015-05	267: 74	931 (28%): 2412 (72%)
2012-12	162:90	1105 (51%): 1055 (49%)	2015-06	280: 80	1208 (31%): 2720 (69%)
2013-01	160: 88	959 (47%): 1101 (53%)	2015-07	293: 83	1388 (32%): 2948 (68%)
2013-02	153:70	575 (39%): 886 (61%)	2015-08	284: 82	1214 (30%): 2839 (70%)
2013-03	196:99	1416 (43%): 1904 (57%)	2015-09	263: 75	1170 (30%): 2682 (70%)
2013-04	195: 92	1324 (38%): 2198 (62%)	2015-10	266: 71	1153 (31%): 2583 (69%)
2013-05	234:112	2304 (42%): 3229 (58%)	2015-11	255: 67	914 (26%): 2620 (74%)
2013-06	218: 85	1317 (33%): 2720 (67%)	2015-12	250: 62	910 (27%): 2452 (73%)
2013-07	208:74	1117 (34%): 2186 (66%)	2016-01	254: 61	1047 (31%): 2341 (69%)
2013-08	210:71	1008 (35%): 1913 (65%)	2016-02	218: 41	343 (20%): 1402 (80%)
2013-09	213:73	1108 (36%): 2003 (64%)	2016-03	259: 36	389 (20%): 1518 (80%)
2013-10	213:75	1076 (33%): 2153 (67%)	2016-04	246: 37	516 (27%): 1427 (73%)
2013-11	225:79	1292 (35%): 2426 (65%)	2016-05	225: 21	162 (16%): 877 (84%)
2013-12	215:81	1102 (33%): 2241 (67%)	2016-06	194: 15	120 (17%): 597 (83%)
2014-01	228: 78	1081 (30%): 2507 (70%)	2016-07	177: 13	83 (16%): 427 (84%)
2014-02	213:74	928 (30%): 2189 (70%)	2016-08	146: 9	54 (14%): 326 (86%)
2014-03	228: 68	852 (27%): 2255 (73%)	2016-09	111:4	34 (16%): 176 (84%)
2014-04	224: 56	564 (23%): 1941 (77%)	2016-10	95: 2	26 (14%): 157 (86%)

5.2   Analysis of Network Evolution at Macro Level

(1) Aggregation coefficient, average shortest path length, and modularity

In addition to the above basic statistical indicators, this paper analyzes the structural characteristics of the newcomer-task network by using aggregation coefficient, average shortest path length and modularity. The aggregation coefficient represents the ratio of the number of triangles to the number of triples in the network; the average shortest path length is the average of the minimum number of edges needed to connect any two nodes; and the modularity mainly measures whether there are many local areas with high edge density in the network (i.e. modularized structure), Louvain algorithm [23] is used to calculate the modularity of the network. Figure 5 shows the evolution of aggregation coefficient (C), average shortest path (L) and modularity (Q) of the newcomer-task network over time windows, where C and L are the ratios of the actual values to the aggregation coefficient and the average shortest path length of the same-scale random network. On the whole, the values of the three indicators have increased and decreased in different degrees. By comparing the changes with the three indicators, we can find that there are three different structural states in the evolution of the newcomer-task network.

The first state appeared in the period of 2011.11–2013.02, and the modularity, average shortest path and aggregation coefficient of newcomer-task network increased in varying degrees (Q → 0.6, L → 1.5, C → 7.5). Because newcomers can't get the guidance and support of experienced users after entering the system, and newcomers exist in different departments, the different tasks of each other make the sparse relationship between them, and the newcomer-task network diagram is in a more decentralized state.

Subsequently, in the period of 2013.02–2016.01, the structure of newcomer-task network shows a relatively stable state: modularity and average shortest path have not changed significantly (Q ≈ 0.5, L ≈ 1.3). It shows that in this stage, while maintaining a highly modular structure, newcomers and all users are gradually interconnected. That is, in the process of newcomers entering and transforming old users, the operation and maintenance of information systems and the updating of functional modules make the process of establishing contact between newcomers and users begin to show a state of mutual integration.

The third state appears in the stage of 2016.01–2016.10, the modularity and shortest path length decrease (Q(0.326 → 0.008), L(1.35 → 0.3)), while aggregation coefficient increases (C(9.94 → 47)). That is to say, the function module of the information system is in accordance with the user's task to a great extent at this stage, which involves the task needs of all users in different departments.

Combining with the existing research on the life cycle of the information system, it is found that the evolution of newcomer-task network can restore the implementation stage of the enterprise information system, that is, it has gone through the new stage, the growing stage and the stable stage of the system successively.

(2) Degree distribution

In order to further analyze the structural characteristics of the newcomer-task network, the degree distribution of the newcomer-task network in the initial month (2011.11) and three different structural breakpoints (2013.02, 2016.01 and 2016.10) are investigated. The corresponding degree distribution maps are drawn in the double logarithmic coordinate system. The power law exponents of the four-time networks are −0.39, −0.46, −0.69 and −0.27, respectively. Figure 6 shows the degree distribution of the newcomer-task network at different times.

It can be seen from the figure that in the double logarithmic coordinate system, the network degree distribution of the four moments shows different distribution trends. In the Fig. 6(1), a small number of nodes have higher degrees of value, and most of the nodes have smaller degrees, that is, after the newcomer enters the system, it is in a state of being scattered with other users. After the enterprise information system was

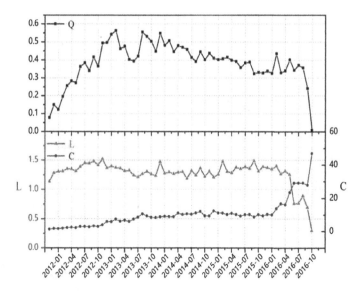

Fig. 5. Newcomer-task network graph C, L, Q value change

launched, a small number of users began to use the information system to ensure the completion of the task, but the contact was not close; In the Fig. 6(2), it can be seen that the degree values of most nodes in this phase are concentrated in a range of activities, that is, newcomers from the information system to the process of infiltration into the enterprise business, Newcomers and other users begin to slowly start to contact due to similar or identical work tasks; In the Fig. 6(3), the network degree distribution shows a linear distribution trend with a tailing phenomenon. At this time, a small number of nodes in the network have higher degrees, while most nodes have lower degrees. The network obeys the power law distribution and satisfies the scale-free characteristics. That is to say, information systems have spread from a few departments to many departments, and centralized databases have been established and systems that can make full use of and manage all kinds of information have been developed. The functional modules are extensively developed, and the relationship between newcomers and other users becomes closer. The business activities of all departments through the information system to complete work tasks are also more frequent; In the Fig. 6(4), the degree distribution appears bipolar phenomenon. Some nodes have smaller degree values and some nodes have higher degree values. It shows that the information system is almost mature, and can meet the requirements of all levels of management (high, middle and grass-roots) in enterprises, thus realizing the management of information resources.

Considering the possible impact of network size on the degree distribution results, the paper uses the power function of Origin software to calculate the power exponent of newcomer task network in each phase and month. From the power exponential results (Table 3), although the newcomer-task network is small, the power exponential results are basically consistent with the analysis results in the case of inconsistent monthly

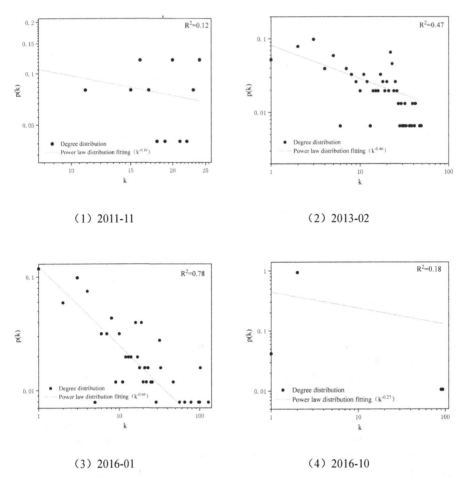

Fig. 6. Distribution of different stages

network size in each stage. In the first stage (2011.11–2013.02), the information system is in a new stage, and a large number of employees begin to learn to use the system to complete their work tasks. Since newcomers just begun to contact the information system, they were not closely related to other users, and their functions are very few and inconsistent, the degree distribution does not show power law distribution. In the second stage (2013.02–2016.01), from Fig. 6(3), it can be seen that the relationship between newcomers and other users of information system has been relatively close, business interpenetration between departments, users gradually rely on information system to complete their tasks. With the continuous improvement of the system, the number of functional modules used has increased more widely. The degree distribution gradually appears power law distribution and keeps relatively stable. In the third stage (2016.01–2016.10), with the gradual improvement of information system function modules and the steady flow of enterprise personnel, the number of newcomers of information system gradually decreased, but still a small number of newcomers entered the system. At this time, the relationship between users in the newcomer-task network begins to turn to extreme, mostly between newcomers and old users, and a few newcomers and newcomers. The power-law distribution of degree distribution begins to disappear.

Combined with (1) (2), through the analysis of aggregation coefficient, average shortest path and modularity of newcomer-task network, this paper divides the evolution process of newcomer-task network into three stages: 2011.11–2013.02, 2013.02–2016.01 and 2016.01–2016.10. At the same time, the degree distribution of the network at the end of the phase is introduced. In order to enhance the reliability of the analysis results, this paper uses Gephi software to visualize the newcomer-task network at each stage. The red node represents the new user, as shown in Fig. 7(1), we can see that when the information system starts to run in the enterprise, newcomers establish links each other through the use of the same functions in the system, but they are scattered. It is found from the database that most users of the information system are distributed in the management department and a small number of users in the basic production department at this stage, and business links are not tight yet; Then in Fig. 7 (2), the first stage, we can find that the newcomers are gradually connected with all users (other newcomers, old users), but there are small groups with inconsistent sizes, which can be interpreted that newcomers from different departments began to enter the system in the process of using system, but because the functional modules are still not perfect, there is a lack of certain links between the departments. It is indicated from the database that the users of the information system are mainly distributed in the management department, the basic production department and the auxiliary production department. At this time, the information system is not popularized in all business links of the enterprise; subsequently, in Fig. 7(3), the newcomer-task network appears to be divided into large and small groups. In the database, the information system function module at this stage is better improved, the departments are more closely connected, and the business activities are more for frequent, at this time, the information system users almost include all departments of the enterprise, such as the management department, the auxiliary production department, the basic production department, the transaction department, and the information implementation department. In Fig. 7(4), it is well verified that two newcomers use the information system function module due to work requirements. It can be found from the figure that the newcomer can use the help of other users to complete the task. It can be better solved when it needs to communicate with other departments.

Table 3. Newcomer task network graph power index per month

Time	R^2	Time	R^2	Time	R^2
2011-11	0.12203	2013-07	0.62279	2015-03	0.77742
2011-12	0.01528	2013-08	0.60509	2015-04	0.82859
2012-01	1.17E−04	2013-09	0.72776	2015-05	0.8371
2012-02	0.00101	2013-10	0.42119	2015-06	0.84343
2012-03	0.0059	2013-11	0.48867	2015-07	0.806
2012-04	5.13E−04	2013-12	0.53238	2015-08	0.7532
2012-05	2.03E−04	2014-01	0.64302	2015-09	0.70033
2012-06	1.59E−04	2014-02	0.58731	2015-10	0.88094
2012-07	0.00372	2014-03	0.56784	2015-11	0.65444
2012-08	5.21E−04	2014-04	0.71734	2015-12	0.78086
2012-09	0.00962	2014-05	0.6818	2016-01	0.77602
2012-10	0.00366	2014-06	0.78383	2016-02	0.73699
2012-11	0.00706	2014-07	0.61625	2016-03	0.79539
2012-12	0.32501	2014-08	0.62503	2016-04	0.77465
2013-01	0.21099	2014-09	0.62466	2016-05	0.89656
2013-02	0.46899	2014-10	0.59647	2016-06	0.66797
2013-03	0.41905	2014-11	0.71426	2016-07	0.6938
2013-04	0.34236	2014-12	0.82172	2016-08	0.62081
2013-05	0.35205	2015-01	0.79984	2016-09	0.34617
2013-06	0.5034	2015-02	0.87937	2016-10	0.18391

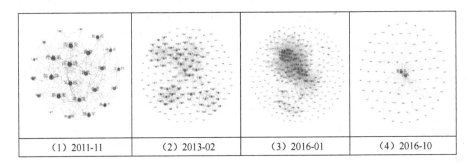

(1) 2011-11	(2) 2013-02	(3) 2016-01	(4) 2016-10

Fig. 7. Newcomer-task network diagram at different stages

6   Conclusion

Taking a shipbuilding enterprise as an example, this paper constructs a newcomer-task network based on the relationship between newcomers and all other users in the system using the same functions. The evolution characteristics of the network structure are analyzed from both micro and macro aspects. The results show that: (1) In the process of system use, newcomers continue to enter, leave or use intermittently,. This result shows that in the process of information system implementation, employees maintain inconsistent attitudes towards information system. Some users believe that the arrival of information system has brought about changes in task structure and business processes. After the first attempt, they give up using this system to complete their tasks. Others think that the functional modules in information system are well integrated with their work. After a short contact, they can improve their work efficiency and continue to use it. In order to keep newcomers using the system continuously, besides strengthening the training of newcomer on the use of information systems, the functional modules of the system need to be improved to further match various business activities; (2) From the connection situation of newcomers, when newcomers first enter the system, because they are not familiar with each other's functional modules, their connections are scattered. However, with the prolongation of the system's use time, the initial newcomers gradually begin to use the relevant functional modules skillfully, and establish close contact with other users. The newcomers who came in later mostly depended on the experienced old users to complete the tasks and improve the work efficiency; (3) According to the analysis results of aggregation coefficient, average shortest path length and modularity, combined with the existing literature about the division of the implementation stage of the enterprise information system, it well match the three states of information system are new period, growing period and stable period. (4) Combining with the distribution map of different stages and the visualization map of newcomer-task network, on the one hand, it can be found that the contact between newcomers and other users is becoming closer and closer from relatively dispersed to small groups to clustering phenomenon with the maturity of enterprise information system; On the other hand, it also shows that the functional modules of enterprise information system have been perfected and gradually spread to various business departments.

The conclusions studied above are helpful to further understand the behavioral characteristics and changing rules of newcomers of information systems. The research results of this paper have a certain reference value for the development of information systems in other manufacturing enterprises or SMEs. At the same time, according to the behavior regulations of newcomers, enterprises can train newcomers in advance to ensure the utilization rate of the system and the efficiency of newcomers. The follow-up work of this paper is mainly carried out in the following three aspects. Firstly, this paper is limited to a single enterprise, not considering the study of multiple enterprises. In order to improve the external effect of the results, we should extend the research to other enterprises and better highlight the theoretical value. Secondly, the change of network location in the process of social network evolution has been paid more attention. In order to better understand the law of newcomers' behavior, the mechanism of network location on their system usage behavior should be further considered. Finally, considering the intersection of functional modules in different departments, the system should be further optimized to reduce the coupling between functional modules.

References

1. Goh, J.M., Gao, G., Agarwal, R.: Evolving work routines: adaptive routinization of information technology in healthcare. Inf. Syst. Res. **22**(3), 565–585 (2011)

2. Hsieh, P.J., Lin, W.S.: Explaining resistance to system usage in the Pharma Cloud: a view of the dual-factor model. Inf. Manag. **55**(1), 51–63 (2018)
3. Dekas, K.: Nooglers to Googlers: applying science and measurement to new hire onboarding. In: Paper presented at the Society for Industrial and Organizational Psychology Conference, Houston, TX (2013)
4. Sykes, T.A., Venkatesh, V., Johnson, J.L.: Enterprise system implementation and employee job performance: understanding the role of advice networks. MIS Q. **38**(1), 51–72 (2014)
5. Kowtha, N.R.: Organizational socialization of newcomers: the role of professional socialization. Int. J. Training Dev. **22**(2), 87–106 (2018)
6. Ellis, A.M., Nifadkar, S.S., Bauer, T.N., et al.: Newcomer adjustment: examining the role of managers' perception of newcomer proactive behavior during organizational socialization. J. Appl. Psychol. **102**(6), 993 (2017)
7. Cai, G.Q., Li, Y.N.: Management information system life cycle system management. Sci. Sci. Manag. S&T **23**(10), 77–80 (2002)
8. Flanagin, A.J., Waldeck, J.H.: Technology use and organizational newcomer socialization. J. Bus. Commun. **41**(2), 137–165 (2004)
9. Kowtha, N.R.: Organizational socialization of newcomers: the role of professional socialization. Int. J. Train. Dev. **22**(2), 87–106 (2018)
10. Roland, N., Frenay, M., Boudrenghien, G.: Towards a better understanding of academic persistence among fresh-men: a qualitative approach. J. Educ. Train. Stud. **4**(12), 175–188 (2016)
11. Nifadkar, S., Tsui, A.S., Ashforth, B.E.: The way you make me feel and behave: supervisor-triggered newcomer affect and approach-avoidance behavior. Acad. Manag. J. **55**(5), 1146–1168 (2012)
12. Lee, R.: Entrepreneurial newcomers: bridging structures and issues of status and power. In: The Social Capital of Entrepreneurial Newcomers, pp. 73–87. Palgrave Macmillan, London (2017)
13. Chiu, M.H.: What does a newcomer digital librarian need to know? Exploring information seeking behavior of newcomer digital librarians in academic libraries during organizational entry. Proc. Am. Soc. Inf. Sci. Technol. **44**(1), 1–6 (2007)
14. Hesketh, E.A., Allan, M.S., Harden, R.M., et al.: New doctors' perceptions of their educational development during their first year of postgraduate training. Med. Teach. **25**(1), 67–76 (2003)
15. McGuire, M.T., Fairbanks, L.A., Cole, S.R., et al.: The ethological study of four psychiatric wards: Behavior changes associated with new staff and new patients. J. Psychiatr. Res. **13**(4), 211–224 (1977)
16. Delobbe, N., Cooper-Thomas, H.D., De Hoe, R.: A new look at the psychological contract during organizational socialization: the role of newcomers' obligations at entry. J. Organ. Behav. **37**(6), 845–867 (2016)
17. Smith, L.G.E., Amiot, C.E., Callan, V.J., et al.: Getting new staff to stay: the mediating role of organizational identification. Br. J. Manag. **23**(1), 45–64 (2012)
18. Chang, L.E., Lin, S.P.: Newcomers' socialization by intranet system. In: 10th IEEE/ACIS International Conference on Computer and Information Science, pp. 239–243. IEEE Press, Sanya (2011)
19. Yin, J., Zhen, Q.Q., Ge, S.L., et al.: Study of effects of functional tasks network's position and relationship on enterprise information system usage. Syst. Eng. Theory Pract. **38**(2), 444–457 (2018)
20. Sykes, T.A., Venkatesh, V., Gosain, S.: Model of acceptance with peer support: a social network perspective to understand employees' system use. MIS Q. **33**(2), 371–393 (2009)

21. Wang, Y., Meister, D.B., Gray, P.H.: Social influence and knowledge management systems use: evidence from panel data. MIS Q. **37**(1), 299–313 (2013)
22. Wiesneth, K.: Evolution, structure and users' attachment behavior in enterprise social networks. In: 2016 49th Hawaii International Conference on System Sciences (HICSS), pp. 2038–2047. IEEE Press, Hawaii (2016)
23. Blondel, V.D., Guillaume, J.L., Lambiotte, R., et al.: Fast unfolding of communities in large networks. J. Stat. Mech: Theory Exp. **2008**(10), 10008 (2008)

Online Knowledge Community Governance Based on Blockchain Token Incentives

Li Zhihong$^{(\boxtimes)}$ and Zhang Jie

South China University of Technology, 381 Wushan Road, Tianhe District,
Guangzhou 510641, Guangdong, China
bmzhihli@scut.edu.cn

Abstract. Blockchain technology as a subversive technology has been widely concerned by academia and industry. The blockchain token economy with the token incentive as its core has also brought about many innovations in incentive mechanism. Based on the characteristics of online knowledge community and the existing problems of incentive mechanism, this paper discusses the innovation of incentive mechanism brought by blockchain technology to online knowledge community. It is found that blockchain tokens can provide a variety of incentive modes, perfect incentive objects, incentive compatibility, and online knowledge community governance through incentives. Also, future research directions have prospected.

Keywords: Online knowledge community · Blockchain · Token economy · Incentive · Governance

1 Introduction

The Internet provides a platform for the general public to participate in knowledge production, and the modes of knowledge production are constantly changing with the development of the Internet. New knowledge production modes, such as user-generated content, shared production, crowdsourcing, collaborative knowledge production and Wikipedia, mode, have a great impact on the knowledge production and management [1]. The main leader of knowledge production has also shifted from enterprise to the public, so "online knowledge community" comprised of "Online social networking" and "knowledge sharing" comes into being, opening a new era of knowledge production. Online knowledge community have created a great deal of social value, business value and knowledge value. However, the existing online knowledge community is facing many problems, such as the low quality of knowledge sharing, the lack of effective protection of intellectual property rights, and the lack of motivation for users to share [2]. How to use new technology and methods to motivate users to share and improve the quality of knowledge sharing is an urgent problem to be solved in the online knowledge community.

The rise of blockchain technology has brought many new projects based on blockchain architecture, and new projects have brought blockchain token economy, a new filed. Bitcoin is the first application of blockchain token economy. Its purpose is to create a decentralized p2p currency payment system [3]. This decentralized monetary

© Springer Nature Singapore Pte Ltd. 2019
J. Chen et al. (Eds.): KSS 2019, CCIS 1103, pp. 64–72, 2019.
https://doi.org/10.1007/978-981-15-1209-4_5

payment system stimulates miners to join the system through mining incentives and transaction cost node incentives and enhances the network effect of the system through miners' activities.

With the development of blockchain technology, token in blockchain has developed from encrypted digital currency to multi-level proof of rights and interests. In recent years, the number and complexity of token in blockchain have increased greatly [4]. In 2015, the establishment of Ethereum marked the entry of blockchain from 1.0 to 2.0. Ethereum is a standardized and universal development platform to promote the application of the bottom technology of blockchain. Users can issue dapp and smart contracts on the platform [5]. Ethereum is a will-designed economic model for the reason that it improves its total market value by pursuing its own interests, but its stable operation is achieved through the design of token incentives. Simultaneously, the size of the blockchain token market has also exceeded 2600 billion dollars. The token economy as the core of the circulation has become one of the frontier fields both academia and industry.

Many new online knowledge communities based on blockchain technology has arisen with the development of blockchain technology, for example, Bihu, Steem and so on, and it's also brought some new incentive mechanisms. However, the effectiveness of these incentive mechanisms still has a lot of room for discussion. So, this study proposes a process for building an incentive comparison between the blockchain token economy and online knowledge community, finding out the incentive mechanism innovation of blockchain token economy to online knowledge community and laying the foundation for the future development of online knowledge community research.

2   Related Work

2.1   Blockchain Token Economy

Before the emergence of the blockchain, the token economy had been widely used. In the 19th century, scholars applied it to the fields of psychiatry and early childhood education. For example, Kelleher and Gollub changed the individual behavior through the token economy system that did not adapt to group behavior [6]. The theoretical basis of the token economy is the enhancement theory, which was first put forward by the behaviorist Skinner [7]. Enhancement theory holds that behavior is a function of behavioral outcomes, and the production of behavior is influenced by behavioral outcomes, so that individual behavior can be changed through specific rewards [7]. Buterin, founder of Ethereum, defined the token economics from the perspective of encrypted digital money. He pointed out that token economics is a practical subject and studied how to effectively manage the production and distribution of goods and services in a decentralized digital economy [8]. Zhang Jun defines the token economy on the blockchain from the perspective of the application of blockchain technology: through incentive mechanism, a new production relationship is constructed, combined with the high-reliability group cooperation mode of blockchain, which can realize a large-scale autonomous' organization system in which the cooperative relationship of maintaining consensus and coexistence of each member in the system can be guaranteed [9].

The blockchain token economy enables programmable incentives through token distortion, which is the core of blockchain. In blockchain token economy, value can be created by and distributed to common users, which used to be created by common user and monopolized by few superintendents [10]. The circulation of token between users promotes the promotion of both the value of token and the value of communities. The value of token is also determined by the quality of users and platforms. Token is the key to the stable operation of blockchain, and it is also an effective tool to restrict the behavior of community users.

2.2 Incentive Mechanism

"Incentive mechanism" is the summation of the structure, mode, relationship and evolution law of the interaction and mutual restriction between the incentive subject system and the incentive object system, in which the incentive subject system uses various incentive means and makes them standardized and relatively immobilized. Means mentioned includes the set of inducing factors, the quality of behavior orientation, the system of behavior range, the system of behavior time and space, and the system of behavior domestication [11]. Among them, induction plays the role of initiating behavior, while the other four systems play the role of guiding, regulating and restricting behavior.

In online knowledge community, incentive mechanism plays an important role. Generally speaking, the incentives for users in online knowledge community can be divided into endogenous incentives and exogenous incentives. Endogenous incentive comes from the pleasure and satisfaction of the individual in the process of behavior, while exogenous incentive is the external reward, such as points, bonuses and so on.

The incentive mechanism is the core of the token economy. The operation of the token economy does not require political or other coordination, it is driven purely by economic incentives, and it is a more open and inclusive economy. In the blockchain token incentive design mechanism, if the incentive mechanism design is unreasonable, it may directly lead to the collapse of the system. Fcoin is an example of the failure of incentive mechanism design which was launched in May 2018 and collapsed in Oct 2018 for its incentive design not durable.

2.3 Online Knowledge Community

Online knowledge Community is a knowledge production platform, comprised of online social and knowledge sharing. Firstly, OKC has a high openness and low entry threshold, a large number of users with cognitive surplus enter the platform to create knowledge using fragmented time, resulting in a large amount of social value; secondly, the integration of OKC platform represented by Wikipedia with traditional industries forms a new industry and produces huge commerce. Finally, OKC itself provides knowledge for other users with knowledge needs through the large amount of knowledge produced by users, creating tremendous knowledge value. However, the existing online knowledge community also has many incentive problems to be solved urgently.

1. The incentive mode is single.

Online knowledge community generally adopts two kinds of incentives: spiritual incentive and material incentive. Spiritual incentive is an invisible incentive to users in spiritual aspect, also known as internal incentive; material incentive is a material incentive to users, also known as external incentive. However, both incentives have limitations. Wikipedia is a good example of spiritual motivation, where users contribute a lot of value, but they do not get corresponding benefits by contributing value. For material incentives, the general material incentives include platform-based integral and direct financial incentives, which can be used for material exchange on the platform. But this kind of material incentive will not last so long that its value is relatively low which will reduce its attractiveness. Integral, which is essentially a liability of the platform, is disadvantageous to the platform itself.

2. Incentive object is imperfect.

Many users have contributed to the online knowledge community, but they have not received good returns. For example, Wikipedia belongs to a non-profit organization whose founder does not obtain benefits through the success of it. Instead, the normal operation of Wikipedia depends on continuous donations.

3. Failure to achieve incentive compatibility.

Some online knowledge communities adopted the integral system. However, the integral that issued by the platform is essentially a liability of the platform. The attribute of the debt is "the formation of past transactions or events of the enterprise, which is expected to lead to the outflow of economic benefits from the current obligations of the enterprise". Therefore, the benefits gained by users through integration are bound to be limited by the platform, and the reward value given by the platform to users is bound to be less than the value created by users. As non-profit organizations, they create a lot of value, but their interests will be protected accordingly. Both of them have not achieved incentive compatibility.

4. Difficulties in governance.

The management and governance of online knowledge community depend on a very small number of members or administrators, while online knowledge community has a great deal of users, knowledge and ineffective errors. Therefore, it is very difficult to rely on a small number of managers to manage, and effective incentive mechanism to promote effective public management is particularly important in online knowledge community.

3 Incentives Innovation on Blockchain Based Online Community

3.1 A Variety of Incentives

Token existed before the advent of the internet. The earliest tokens were tokens for early childhood education and incentives for psychiatric treatment. In addition, Casio for settlement in casinos was also token. In the era of Internet, token appeared in IBM's Token Ring Network protocol, representing holders have privilege of visiting. After the blockchain came into being, token refers to the encrypted digital currency on the blockchain. After it's a simple currency, smart contract came into being, the function of token and the number of tokens has increased [4], so token's attributes also showed diversity. Up to now, there is no generally accepted definition of token in academia and business circles. The reason for this problem can be attributed to the fact that block-chains and token have been developing and the number and attributes of token are diverse [13]. Blow shows some incentive from on token economy (Table 1).

Table 1. Different types of token incentive

Form	Content	Example
Coin	Value storage and exchange media	Block Award
Privileges	Decision-making or other Rights	Transaction costs
Rewards	Tools to encourage participants to perform well	Steemit Proof-of-brain
Penalties	Negative Incentives	Removal of User Rewards

3.2 Incentive for Every Contributors

When a new project or company was founded, the new products he brought were rarely accepted by the general users. One of the important reasons is that the new products lack the network effect [15]. Therefore, it is particularly important to encourage users to use the platform in the early stage of the project of the token economy. ICO, airdrop and reducing the difficulty of mining are the three ways to motivate early users to join the blockchain network.

Project initiation team, which proposes project solutions and implements them. The sponsor of the project has a vital impact on the success of the project, so the incentive of the project sponsor team is particularly important. Usually the team will reduce the difficulty of mining through pre-mining or direct token rewards. Project sponsor team's awards are closely related to project success [15]. Therefore, project team will work harder to contribute to the project and achieve incentive compatibility.

3.3 Incentive Compatibility

Hurwiez first put forward the incentive compatibility theory. The incentive compati-bility principle simply means that the individual rationality can be compatible with the collective rationality, that is, the policy can achieve the effect of subjectivity for oneself and objectivity for others [16].

Firstly, in order to attract users, online knowledge communities will adopt certain subsidies or privileges when they are established, such as invitation system. However, with the growth of users, these users are not rewarded for the benefits of platform growth. In the incentive of token economy, there will be additional token incentives for early users to join the platform, which on one hand is a way to attract users, on the other can enhance the value of token. Although the number of token incentives will decrease in the project maturity period, the token obtained by users will also have higher value. The incentive effect is durable (Table 2).

Table 2. Value comparison of different incentive

	Token incentive	OKC incentive
Incentive from	Token	Integral or spiritual motivation
Value	Increase with the number of users	No real value or a liability of the platform

Secondly, some online knowledge communities adopt the integral system. However, the integral issued by the platform is essentially a liability of the platform. Liabilities are formed by the past transactions or events of the enterprise, which is expected to lead to the outflow of economic benefits from the current obligations of the enterprise. The biggest difference between the blockchain token and the traditional integral is that it does not belong to the platform's liability. In addition, the total value of the token economy is closely related to the value of the token, so the holder of the token belongs to the users of the platform, shareholders and employees of the platform.

In addition, non-profit platforms such as Wikipedia, which create tremendous value for society, often fail to gain value themselves, and often have to ask for donations to finance the sustainable development of projects [17]. Blockchain token provide a new way to finance open source projects and support sustainable development. Using blockchain technology, open source projects can be funded by issuing blockchain tokens rather than by donations. By issuing blockchain tokens, open source projects can capture some of the value they create and provide them with financial resources to support their sustainable development [15].

Finally, the traditional platform gains a large number of users through free products to form monopolies and barriers, and then profits through advertising and value-added services on this basis. After monopoly, a highly centralized organization has been formed, super-high barriers have been established, and the marginal effect of the Internet has tended to zero. The ideal situation is to provide everyone with what they want by precise matching and realize super-high profit rate. But there is also a paradox: the value of Internet enterprises increases exponentially due to the increase in the number of users, but users themselves do not benefit directly from it. Token economy is totally different, whose value can be spread in the blockchain network so that business logic will become how to create value, how to establish greater benefits. We need to think more about how to produce an excellent token economic model, so that partners, suppliers, customers, and even competitors contribute more for the whole ecosystem.

3.4   Incentive Based User-Self Governance

For the online knowledge community, the quality of knowledge shared by users and the bad behavior of users need to be addressed. On traditional platforms, governance may be done by a few central managers. However, the traceability and de-centralization of blockchains, combined with the benefit distribution and IT, can realize the autonomy of online knowledge community. The simple definition of IT governance is to maximize the interests of all parties involved in the information process. In IT governance, the most critical three elements are incentive, decision-making and accountability [19]. However, blockchain token can realize multiple value representations. Therefore, decision-making power can also be realized through token incentives (Fig. 1).

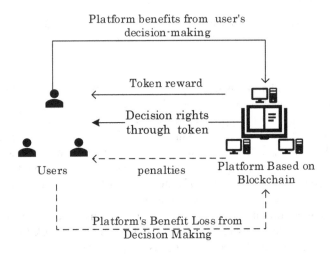

Fig. 1. Governance based on Blockchain

Another potential incentive is that the traceability of the blockchain makes the user's behavior accountable. Decision-making power is the decision-maker's power to choose, control and dominate the activities in the decision-making system, which involves the interests of the decision-makers. Therefore, accountable agents must address actions taken and consequences incurred [19]. The accountability of management is usually executed, designated and implemented by contract or legal framework. Blockchain smart contract provides basic conditions for accountability. Blockchain traceability provides an effective tool for consequences traceability actions. Finally, rewards and penalties on blockchains can reinforce effective actions and punish actions with adverse consequences [20].

4   Conclusions

The main motivation of this article is to explore an incentive innovation that blockchain has bring to online knowledge community. The object of study is the online knowledge community, and the main research question is concerned about what blockchain technology brings about the innovation of incentive mechanism for online knowledge community. Firstly, this study aims at the single incentive mode and imperfect incentive object of existing incentive mechanism for online knowledge community.

The existing problems such as incentive compatibility and user management difficulties are sorted out. Secondly, this study analyses the multiplicity of incentive modes of blockchain from the perspective of the multiplicity of blockchain tokens. This multiplicity of incentives can bring more comprehensive incentive modes to online knowledge community. Thirdly, from the perspective of the design of the token economy, this paper discusses blockchain token incentives. In the knowledge community, incentives for all contributors and incentive compatibility can be realized. Finally, this study explores the possibility of blockchains in the governance of online knowledge communities. Through blockchains token incentives, online knowledge users can be given decision-making power, while through blockchains traceability, behavior accountability can be achieved. In this way, rewards for effective behavior and negative incentives for behavior with adverse consequences can be realized.

5 Discussion

The limitations of this study are as follows. Firstly, blockchain technology has a wide range of applications, and different incentive mechanisms should be provided in different application scenarios. However, the conclusions from the online knowledge community selected in this study may not be applicable to other industries. In addition, this paper uses qualitative method to study the incentive mechanism innovation brought by blockchain technology to online knowledge community, but the user behavior of online knowledge community has always been a complex problem, and the incentive effect of blockchain online knowledge community needs further empirical research.

Following this promising work, future research should address the design of token incentive mechanism for online knowledge community based on blockchain to promote the quality of online knowledge sharing and ecological formation of users. In addition, the incentive effect of online community users based on blockchain token still needs further empirical research to explore, so as to form a more universal incentive theory for online knowledge users.

References

1. Zheng-kui, L., Feng-jun, L., Na, Z.: Research review on intrinsic mechanism of group collaboration in online knowledge community. Inf. Sci. **37**(06), 170–177 (2019)
2. Halfaker, A., Geiger, R.S., Morgan, J.T., Riedl, J.: The rise and decline of an open collaboration system: How Wikipedia's reaction to popularity is causing its decline. Am. Behav. Sci. **57**(5), 664–688 (2013)
3. Nakamoto, S.: Bitcoin: a peer-to-peer electronic cash system (2008)
4. Oliveira, L., Zavolokina, L., Bauer, I., Schwabe, G.: To token or not to token: tools for understanding blockchain tokens (2018)
5. Wood, G.: Ethereum: a secure decentralised generalised transaction ledger. Ethereum Project Yellow Paper **151**(2014), 1–32 (2014)
6. Kelleher, R.T., Gollub, L.R.: A review of positive conditioned reinforcement 1. J. Exp. Anal. Behav. **5**(S4), 543–597 (1962)

7. Chomsky, N.: Review of verbal behavior by BF Skinner (2003)
8. Buterin, V.: A next-generation smart contract and decentralized application platform. White Paper **3**, 37 (2014)
9. Zhang, J., Wang, F.Y., Chen, S.: Token economics in energy systems: concept, functionality and applications. arXiv preprint arXiv:1808.01261 (2018)
10. Kim, M., Chung, J.: Sustainable growth and token economy design: the case of steemit. Sustainability **11**(1), 167 (2019)
11. Zhang, K., Antonopoulos, N., Mahmood, Z.: A review of incentive mechanism in peer-to-peer systems. In: 2009 First International Conference on Advances in P2P Systems, pp. 45–50. IEEE (2009)
12. FCoin Crypto Exchange Draws Fire for Controversial Business Model. https://www.coindesk.com/new-crypto-exchange-draws-fire-over-controversial-business-model. Accessed 21 July 2019
13. The Token Classification Framework: A multi-dimensional tool for understanding and classifying crypto tokens. http://www.untitled-inc.com/the-token-classification-framework-a-multi-dimensional-tool-for-understanding-and-classifying-crypto-tokens/. Accessed 21 July 2019
14. Lipusch, N.: Initial Coin Offerings–A Paradigm Shift in Funding Disruptive Innovation. Available at SSRN 3148181(2018)
15. Chen, Y.: Blockchain tokens and the potential democratization of entrepreneurship and innovation. Bus. Horiz. **61**(4), 567–575 (2018)
16. Hurwicz, L.: The design of mechanisms for resource allocation. Am. Econ. Rev. **63**(2), 1–30 (1973)
17. Glott, R., Schmidt, P., Ghosh, R.: Wikipedia survey–overview of results. United Nations University: Collaborative Creativity Group, pp. 1158–1178 (2010)
18. Atzori, M.: Blockchain technology and decentralized governance: is the state still necessary? SSRN 2709713 (2015)
19. Beck, R., Müller-Bloch, C., King, J.L.: Governance in the blockchain economy: a framework and research agenda. J. Assoc. Inf. Syst. **19**(10), 1020–1034 (2018)
20. Weitzner, D.J., Abelson, H., Berners-Lee, T., Feigenbaum, J., Hendler, J., Sussman, G.J.: Information accountability. Commun. ACM **51**(6), 82 (2008)

The Formation and Effect of Affect in Knowledge Intensive Team: A Dynamic Computational Model

Xin Yue, Yanzhong Dang, Deqiang Hu$^{(\boxtimes)}$, and Jiangning Wu

Institute of System Engineering, Dalian University of Technology,
Dalian 116024, China
deqianghu@mail.dlut.edu.cn

Abstract. This paper's purpose is to investigate the formation process of affect, the mechanisms by which it functions, and the dynamic characteristics of its influence on team performance. Toward this end, we present a computational experiment based on agent-based modeling. In particular, the modeling deeply penetrates internal psychological activities. Two computational experiments are conducted under different internal and external conditions for the team, yielding the following results. The process of affect formation is influenced by not only difficulty but also order of the task. Processing task from simple to difficult improves the formation of positive affect and facilitates team performance. While processing task from difficult to simple leads to the formation of negative affect and obstructs team performance. This research extends prior findings by examining the dynamic interplay of the determinants of affect over time and the study method presented herein are appropriate for other studies focusing on psychological effects on team, laying the foundations for new ideas for studying team building and team development.

Keywords: Affect · Team performance · Computational experiment · Interpersonal knowledge interaction · Computational model

1 Introduction

Researchers from management and related fields have agreed that affect has important organizational and interpersonal consequences, especially on performance [1]. However, given the apparent relationship between affect and team performance, it is important to understand how affect is nurtured and eroded as parties interact. Few studies have examined affect between parties over time. Particularly in knowledge-intensive team (hereafter just "team") this complex social context [2], team members accomplish their tasks through interpersonal knowledge interaction that makes knowledge transferred, exchanged and shared between members [3]. In psychology, affect is special form of reaction to objective reality. It is the subjective experience and attitude that whether the objective things meet the need of individual. Affect thus is not only an important factor on interpersonal knowledge interaction, but also a significant consequence of this interaction. Therefore, revealing how affect forms and varies in

© Springer Nature Singapore Pte Ltd. 2019
J. Chen et al. (Eds.): KSS 2019, CCIS 1103, pp. 73–86, 2019.
https://doi.org/10.1007/978-981-15-1209-4_6

interpersonal knowledge interaction can make us better understanding the relationship between affect and team performance in team development perspective and strengthen the effectiveness of team management.

After a interpersonal knowledge interaction, a member naturally makes an appraisal of interaction results and interaction object. If the interaction results satisfy the member's need, the member forms positive attitude experience and arouses positive emotion to the interaction object[4]; otherwise the member forms negative attitude experience and arouses negative emotion to the interaction object. Though emotion is transient and fluctuant, affect is relatively permanent and stable [5]. Affect, thus, accumulates with the growing and diminishing of emotion and influences following interpersonal knowledge interaction [6], and ultimately team performance [7]. Therefore, this study pays more attention to affect dynamic rather than some specific affective state such as happy, angry and sad, and also focuses on dynamic relationship between affect and team performance.

The existing method about affect, interpersonal knowledge interaction and team performance is mainly empirical study and the researchers get fruitful research results. For example, positive affect can motivate members' work enthusiasm [8], improve members' interaction efficiency [15], reinforce links between members [9], create a harmonious working atmosphere [16] etc. Thus, positive affect has a positive influence on team performance [10]. While negative affect may reduce the enthusiasm of members [11], impact the effect of member communication and cooperation [18], impede the formation of team cohesion, destroy innovation environment, etc. Thus, negative affect has a negative influence on team performance [12, 13]. However, with empirical method, it is challenging to obtain individual-level data on interaction and affect variation. As a result, empirical method is not quite suitable to study affect's formation process and its dynamic effect on team performance. Even thought, the conclusions of the existing empirical research [14, 15] can still be used as evidence for the methods and models in this paper.

Computational experiment offers a powerful tool for deciphering complex social phenomenon that emerges through agents' interaction, specific action and dynamic decision and exploring the aggregate and long-run findings [16]. Thus, computational experiment method is suitable to study affect's formation process. However, the existing computational experiment researches have paid much attention on agents' individual behavior rules and are lack of consideration about the psychological activities in modeling [17]. So, in studying affect's effect on team performance through interpersonal knowledge interaction, we need to fully consider the interaction of multiple mechanisms of psychology and behavior among members of the real team.

This study adopts bottom-up modeling method based on ABM and the modeling especially focuses on psychological activities. We also establish corresponding psychological and behavioral mechanisms and test our models and results using existing empirical research conclusions. Thereupon, we conducted different computational experiment through different task difficulties and orders, supporting our purpose to explore affect's formation process, and to analyze and revel the dynamic relationship between affect and team performance in team development context.

2   Modeling

2.1   Research Framework

This study simulates how team develops during task processing process. Sometimes, members' knowledge can't satisfy the need to accomplish the task allocated all by themselves. So they need to obtain knowledge by making interpersonal knowledge interaction with other members. And to what extent the tasks have been finished is evaluated by team performance. The team concept model is shown in Fig. 1. Individual-level affect's generation, accumulation and effect process springs from interpersonal knowledge interaction and individual-level affect emerges into team-level affect [18, 19], The different tasks lead to different interpersonal knowledge interaction and different formation of affect, which may create different team affect and team performance. Therefore, we set tasks, as critical driven force of team development, as important experiment condition [20]. We have conducted various computational experiment under different task difficulties and orders. Our primary research objectives are to explain how affect operates at the individual level of analysis, how it is related to interpersonal knowledge interaction, and, particularly, the mechanisms by which this inherently individual-level phenomenon dynamically translates into a team-level outcome: team affect and team performance. Modeling may be especially suitable because empirical work in this area suffers from severe data limitations. It is challenging to obtain individual-level data on knowledge and transfer and extremely difficult to gain access to sites for conducting field experiments or replicating past results.

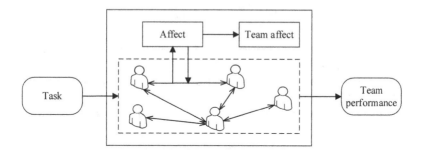

Fig. 1. Team model

2.2   Affect Model

The formation of affect comes out with three processes [4]: (1) generation of emotion, (2) accumulation of emotion to affect, (3) effect of affect. This three processes functions in different conditions. When agent (A_i) acts as a knowledge demander, its interpersonal knowledge interaction involves the generation and accumulation of its affect to others. While agent (A_j) acts as a knowledge supplier, its interpersonal knowledge interaction involves the effect of its affect to others. These three affect processes occurs alternately with the development of interpersonal knowledge interaction. Therefore, affect forms in interpersonal knowledge interaction and, in turn, affects knowledge transfer therein, ultimately influencing team performance. The next content focus on the modeling of this three affect processes.

Generation of Emotion. Emotion is the inner feeling that whether objective things satisfy the subjective needs [21]. Emotion is closely related to the needs of the subject, so the needs, as the intermediaries, act as the manifestation. The basic element that determines the state of emotion is to what extent the needs are satisfied. The satisfaction state of both obvious needs and potential needs determine the state of emotion. Usually, the satisfaction of needs cause positive feelings and the dissatisfaction of needs cause the negative feelings. Positive feelings leads to positive emotions and negative feelings leads to negative emotions.

In interpersonal knowledge interaction, after A_i made a interaction with A_j, A_i's satisfaction to A_j generates by evaluating and analyzing the interaction results. According to the S-O-R model [22], satisfaction as stimulation (S), under the psychological internal analysis and evaluation of organism (O), leads to the generation of emotion that denotes as reaction (R). Soon afterwards, the intensity of emotion reaches the maximum value. Then individual gradually calms and his intensity of emotion gradually weakens over time, eventually subsides [7].

In nth interpersonal knowledge interaction, if A_i gets knowledge from A_j, then A_i evaluates this interaction and generates satisfaction of A_j. The satisfaction means to what extent the A_i's knowledge need is satisfied by A_j. This satisfaction is denoted as $S_n(A_i, A_j)$, defined as:

$$S_n\left(A_i, A_j\right) = \frac{K_n^{\text{get}}\left(A_i, A_j\right)}{K_n^{\text{need}}\left(A_i\right)} \tag{1}$$

Where $K_n^{\text{get}}(A_i, A_j)$ denotes the amount of knowledge that A_i gets from A_j, and $K_n^{\text{need}}(A_i)$ denotes the amount of knowledge that A_i needs

The emotion generates when the knowledge need is satisfied with some certain degree. The higher degree is the satisfaction, the higher intensity is positive emotion, and vice versa. A_i's emotion to A_j is denoted as $e_n(A_i, A_j)$, defined as:

$$e_n\left(A_i, A_j\right) = \begin{cases} \log_{0.5}^{2-2S_n\left(A_i,A_j\right)}, 0 \leq S_n\left(A_i, A_j\right) < 0.5 \\ \log_2^{2S_n\left(A_i,A_j\right)}, \ 0.5 \leq S_n\left(A_i, A_j\right) \leq 1 \end{cases} \tag{2}$$

Accumulation of Emotion to Affect. Although there are different types of affect, in this article an affect can be "good" or "bad", depending on the team member's evaluation of the interpersonal knowledge interaction. An affect is the precipitation and accumulation of emotions. However, it is relatively perdurable and stable, compared to the transience and fluctuating nature of an emotion [23]. The affect's generation process is the accumulation of several emotions. That is, a newly generating emotion builds and combines to form an affect that has already combined with a previous emotion [7]. Take positive affect's accumulation process for example. In real life, the accumulation of an affect is a complex process. The incremental speed of an affect is from slow to fast, and then it becomes stable and slows into a trend. At first, the individual is not familiar with the object they are interacting with and, initially, the incremental speed of the affect is relatively slow. With the increasing frequency of interactions, the affect is rapidly heated up. At last, a stable affect has been built between the individual and the object they are interacting with. At this point, the speed increment calms down and the intensity of the affect is infinitely close to the upper limit. The accumulation of negative emotions corresponds to the accumulation of positive emotions, but it is only a process of gradual reduction.

A_i's affect to A_j after nth interaction is denoted as $Q_n(A_i, A_j)$, defined as:

$$Q_n(A_i, A_j) = \left(1 - \frac{1}{n}\right) Q_{n-1}(A_i, A_j) + \frac{1}{n}\left[e_n(A_i, A_j) \times \frac{1}{1 + e^{(-n)}}\right] \tag{3}$$

Effect of Affect. Affect has an indispensable influence on people's decisions and behaviors [24]. Affect is not only an important part of the driven force of psychological activities, but also an important influencing factor of the regulation mechanism in behavior. In general, positive affect can increase the enthusiasm, promote the success and improve the effectiveness of behavioral activities. While negative affect decrease the enthusiasm, obstruct the achievement and reduce the effectiveness of behavioral activities. Therefore, positive affect makes people closer, which can be conducive to the communication and cooperation of members, thus promoting knowledge transfer. While negative affect may initiate harmful team interaction processes that inhibit knowledge exchange and sharing.

In interpersonal knowledge interaction, the more affect A_i places in A_j, the more knowledge A_i is willing to transfer to A_j, and vice versa. When A_i gives K_g type of knowledge to A_j, the knowledge amount $K_n^{\text{give}}(A_i, A_j)$ defined as follows:

$$K_n^{\text{give}}(A_i, A_j) = \left[K_g(A_i) - K_g(A_j)\right] \frac{1 + Q_{n-1}(A_i, A_j)}{2} \tag{4}$$

Our modeling illustrates how affect generates and accumulates through the psychological activities from dynamic perspective which is lack in other existing models. And we also consider the circle of affect's generation and effect which is challenging to the study but is closer to the real situation.

3 Computational Experiment

Based on our model, we conduct our experiment by different settings, then analyze experiment results to illustrate how affect forms in team development and how affect influences team performance dynamically. In turn, it may provide decision support for team managers to develop correct management strategies and improve the effectiveness of affect regulation. Thus, we design following two computational experiment.

(1) We simulate different task performing process under different affect degree to verify the rationality of our model, then we examine whether affect's formation is influenced by task.

(2) We conduct our experiment on the subject of this article. We examine how affect forms and how affect influences team performance dynamically under different task difficulties and orders.

3.1   Experiment Input

Task. As an input to the team system, task is an important driving force for team development. A team can undertake one or more batches of tasks in a certain period of time. T(team) defined as:

$$T(\text{team})=\{T_1, T_2, \cdots, T_h, \cdots, T_p\} \tag{5}$$

Where T_h represents a batch of tasks, p represents the number of batches. T_h comprises several sub-tasks that are relatively independent, defined as:

$$T_h = \{ST_i | i = 1, 2, \cdots, q\} \tag{6}$$

Where ST_i represents a sub-task in T_h. Each sub-task can have multiple attributes, and this paper essential considers the knowledge attribute. A task's knowledge attribute determines the required types and level of knowledge.

For this study's purposes, experience, skill, and ability are collectively called knowledge, which is the basic attribute for both of the member and task. The space K of d dimensions denotes the knowledge space, where each dimension represents one type of knowledge. K is defined as:

$$K = \{K_1, K_2, \cdots, K_d\} \tag{7}$$

Sub-task ST_i's knowledge attribute is a vector in K. It is denoted as $k(ST_i)$, defined as:

$$k(ST_i) = \left(k_1(ST_i), k_2(ST_i), \cdots, k_j(ST_i), \cdots, k_d(ST_i)\right) \tag{8}$$

Let $k_j(ST_i) \in [0,1], j = 1, 2, \ldots, q$ denote the amount of ST_i's K_j type of knowledge. There exists at least one j for which $k_j(ST_i) > 0$.

Task Difficulty. If any type of knowledge K_j of any subtask ST_i of T_h satisfies $k_j(ST_i) \in [x_1, x_2]$, then task knowledge domain $[x_1, x_2]$ denotes the task difficulty of T_h.

Team Ability. If any team member A_i's any type of knowledge K_j satisfies $k_j(A_i) \in [y_1, y_2]$, then task knowledge domain $[y_1, y_2]$ denotes the team ability.

3.2   Experiment Output

Team Affect. Team affect Q(team) denotes the average affect of the team members that have affective relationships, defined as:

$$Q(\text{team}) = \frac{1}{m'} \sum_{i=1}^{m'} \sum_{j=1, j \neq i}^{m'} Q(A_i, A_j) \tag{9}$$

Where $m' = \sum_{i=1}^{n} \sum_{j=1}^{n} v_{ij}$ is the total number of affect relationships that exist between members. Let $v_{ij} = 0$ denote there exist affective relationship between A_i and A_j. Let $v_{ij} = 1$ denote there doesn't exist affective relationship between A_i and A_j.

Team Initial Affect. If any team member A_i's affect to A_j satisfies $Q(A_i, A_j) = z$, then z denotes team initial affect.

Team Performance. Team performance is the degree to which tasks are performed within the allotted resources, conditions and circumstances and a measure and feedback on the achievement of goals and the achievement efficiency [25]. In this paper, team performance is measured by task-completion rate.

Task-completion rate is the degree to which a task has been completed. It is denoted as $CR(T)$, defined as:

$$CR(T) = \frac{N^{\text{finish}}(T)}{N^{\text{task}}(T)} \tag{10}$$

Where $N^{\text{finish}}(T)$ is the number of tasks that the team has completed, and $N^{\text{task}}(T)$ is the number of tasks that the team has accepted. Due to different knowledge levels and the inhomogeneity of knowledge distribution between members, the task may not be completed within the allotted time.

3.3   Experiment Parameters

The parameters of two computational experiment are shown in Table 1. Experiment 1 aims at verifying of our models. Thus, the team in experiment is not in dynamically developing context, which means that, when each time the team completes the task, the team returns to the initialized state and then performs the next task. Experiment 2 focuses on the subject of this article. Thus, the team in experiment is in dynamically developing context, which means that, when each time the team completes the task, the team continues on the basis of the last task performed and then performs the next task. We use Boolean parameter td to denote whether team is dynamically developing. Let td = false denote that the team is not dynamically developing. And let td = true denote that the team is dynamically developing. Other parameters are mentioned in modeling section and different values are set according to different experiments.

3.4   Experiment 1

Design of Experiment 1. Experiment 1 includes two sub experiment.

(1) Team initial affect is independent variable and team performance is dependent variable. Task difficulty and team ability are control variable.

Table 1. Parameters of the simulation

Parameter	Experiment 1		Experiment 2	Implication
	(1)	(2)		
d	10			knowledge dimension
m	10			the number of team members
p	50		60	the number of tasks
q	10			the number of subtasks
$[x_1,x_2]$	(0,0.02], (0.02,0.04], ..., (0.98,1] (total 50 kinds)		Simple:[0,0.2], Moderate:[0.2,0.5], Difficult:[0.5,1]	task difficulty
$[y_1,y_2]$	[0,01.0.5], [0.02,0.51], ..., [0.5,1] (total 50 kinds)		[0.25,0.75]	team ability
z	−1, −0.5, 0, 0.5, 1	0		team initial affect
td	false		true	whether team develops

(2) Task difficulty is independent variable and team affect is dependent variable. Team ability is control variable.

The experiment 1's parameters setting are shown in Table 1.

Results of Experiment 1

Each three-dimensional diagram in Fig. 2 denotes a team with a kind of team initial affect. X axis denotes task difficulty. Y axis denotes team ability. Z axis denotes team performance that defined as task-completion rate. When task difficulty is fixed, task-completion rate doesn't have appreciable change with variation of team ability. And when team ability is fixed, task-completion rate has appreciable change with variation of task difficulty. This result indicates that task difficulty and team ability have a combined action on team performance.

Thereupon, we take one team ($[y_1, y_2] \in [0.2,0.7]$) for example. As shown in Fig. 3, different curves denotes different team initial affect team. And different variation trend of five curves suggests that affect has obvious significance on team performance. In addition, the team with higher team initial affect has slower decline and the smaller amplitude of task task-completion rate. And vice versa.

As shown in Fig. 4, X axis denotes task difficulty. Y axis denotes team ability. Z axis denotes team performance that defined as task-completion rate. When task difficulty is fixed, team affect doesn't have appreciable change with variation of team ability. And when team ability is fixed, team affect has appreciable change with variation of task difficulty. This result indicates that task difficulty, compared to team ability, has a greater impact on the generation of team affect.

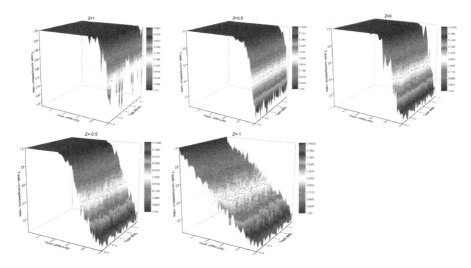

Fig. 2. Team performance across different task difficulty and team ability under different affect (multiple team)

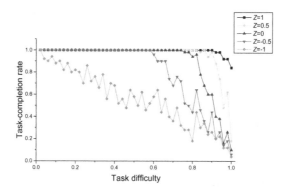

Fig. 3. Team Performance across different task difficulty under different affect (single team)

Thereupon, we take one team ($[y_1, y_2] \in [0.2,0.7]$) for example. As shown in Fig. 5, team affect and task difficulty show inverted U-shaped relationship. Team affect first increases with the difficulty of the task, and reach the maximum when the task difficulty is about 0.2; then decreases with the increase of the task difficulty, the team affect drops to 0 when the task difficulty is about 0.5; then the team affect changes to negative value, and further decreases as the difficulty of the task increases. This result suggests that different task difficulty has different effect on the generation of team affect.

Findings of Experiment 1. The results of experiment 1 suggest that the team with higher team initial affect has better team performance under different team ability. The results are consistent with the existing research [2, 14, 20], which proves the rationality of the model. And we also find that the generation of team affect has close relationship with task difficulty and team ability.

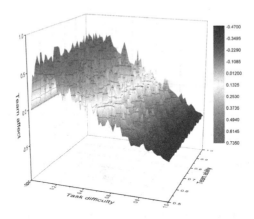

Fig. 4. Team affect across different task difficulty and team ability

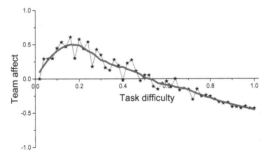

Fig. 5. Team affect across different task difficulty

3.5   Computational Experiment 2

Design of Experiment 2
According to the results of experiment 1, the generation of team affect of different team, under different team ability, all have similar variable trend. Thus, experiment 2 takes one team for example, and set $[y_1, y_2] \in [0.25, 0.75]$ without loss of generality. As shown in Fig. 5, based on the inflection points, the task difficulty is divided into three levels: $[0, 0.2]$(simple task $\alpha 1$); $[0.2, 0.5]$(moderate task $\alpha 2$); $[0.5, 1]$(difficult task $\alpha 3$). And based on this three levels the task orders are divided into six types: (1) $\alpha 123$: $\alpha 1 \rightarrow \alpha 2 \rightarrow \alpha 3$, (2) $\alpha 132$: $\alpha 1 \rightarrow \alpha 3 \rightarrow \alpha 2$, (3) $\alpha 213$: $\alpha 2 \rightarrow \alpha 1 \rightarrow \alpha 3$, (4) $\alpha 231$: $\alpha 2 \rightarrow \alpha 3 \rightarrow \alpha 1$, (5) $\alpha 312$: $\alpha 3 \rightarrow \alpha 1 \rightarrow \alpha 2$, (6) $\alpha 321$: $\alpha 3 \rightarrow \alpha 2 \rightarrow \alpha 1$. The experiment 2's parameters setting are shown in Table 1.

Results of Experiment 2

As shown in Fig. 6, although the team performs tasks of exactly the same three kinds of difficulty every time, the process of affect formation is different due to the different order of tasks. When the team affect tends to be stable, the team affect order from large to small is $\alpha123 > \alpha213 > \alpha231 > \alpha132 > \alpha312 > \alpha321$. The team affect is largest when it processes the tasks with the difficulty order from simple to difficult. While the team affect is smallest when it processes the tasks with the difficulty order from difficult

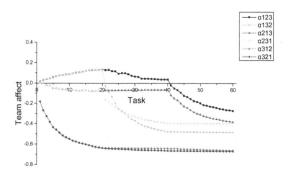

Fig. 6. The formation of team affect across different task orders

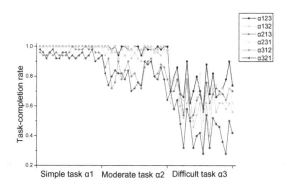

Fig. 7. Team performance across different task difficult under difficult task orders

to simple. In addition, all the curves drop rapidly at some stage, and this stage is the one in which the team performs difficult tasks. Comparing with two curves of $\alpha132$ and $\alpha231$, we also find that when the team starts to perform the task, the team affect of $\alpha132$ is higher than that of $\alpha231$; but when the team affect is stable, the team affect of $\alpha132$ is lower than that $\alpha231$. Because the improvement of team affect is a slow process. Although simple task facilities the formation of team affect, team affect takes time to develop. While the reduction of team affect is very rapid, and it is likely that the team affect will be disrupted due to certain unexpected situations. Therefore, when the team has an affect crisis, it must be remedied in time.

As shown in Fig. 7, although the team performs tasks of exactly the same three kinds of difficulty every time, the variation of team performance is different due to the different orders of tasks. As shown in Fig. 8, the task order has different impact for different task difficulties. The team performance under different task orders has minor difference with each other when the team processes simple tasks. And this difference is gradually

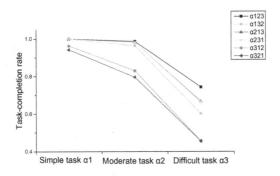

Fig. 8. Average team performance across different task difficult under difficult task orders

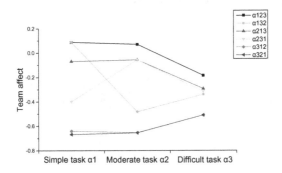

Fig. 9. Average team affect across different task difficult under difficult task orders

expanding with the increase of task difficulty. Comparing Fig. 8 with Fig. 9, it demonstrates that the task order of task-completion rate corresponds to the task order of team affective. In other words, the team has both largest team affect and best team performance when team processes the tasks as the task order of α123; while the team has both smallest team affect and worst team performance when team processes the tasks as the task order of α321. Because different task order leads to different formation processes of team affect which in turn influences team performance.

Findings of Experiment 2. The results of experiment 2 shows that the process of affect formation is influenced by not only difficulty but also order of the task. And different affect formation processes have different influence on team performance. Processing task from simple to difficult improves the formation of positive affect and thus facilitates team performance. While processing task from difficult to simple leads to the formation of negative affect and thus obstructs team performance.

4 Conclusion

In this paper, we have examined interpersonal knowledge interaction to explicate formation process of affect and the connection between affect and team performance in knowledge intensive team context. We have established computational model based on ABM, and have tested our model and method through the existing empirical results. Our results suggest that affect has an indispensable influence on team performance in the dynamic process of team development. In addition, the formation of affect is influenced by the difficulty and order of the task. Processing task from simple to difficult is conducive to the development of positive affect and then promotes team performance. While processing task from difficult to simple is harmful to the development of positive affect and then restrains team performance.

Our research method is particularly suitable for tracking dynamic process of psychological activity in team development, provides a new research approach for the effect of affect on team performance and team development, and has great potential for researching related issues about team management and team building. Although the experimental results are only analyzed for the process of affect intensity, we will analyze the formation of affect network between the team members in the future. At the same time, we hope that this article will arouse the resonance of scholars in this area.

Acknowledgment. This work was partly supported by the National Natural Science Foundation of China under Grant No. 71471028 and No. 71871041.

References

1. Bankes, S., Lempert, R., Popper, S.: Making computational social science effective: epistemology, methodology, and technology. Soc. Sci. Comput. Rev. **20**(4), 377–388 (2002)
2. Barsade, S.G., Gibson, D.E.: Why does affect matter in organizations? Acad. Manage. Perspect. **21**(1), 36–59 (2007)
3. Brannick, M.T., Roach, R.M., Salas, E.: Understanding team performance: a multimethod study. Hum. Perform. **6**(4), 287–308 (1993)
4. Brief, A.P., Weiss, H.M.: Organizational behavior: affect in the workplace. Annu. Rev. Psychol. **53**(1), 279 (2002)
5. Côté, S.: Affect and performance in organizational settings. Curr. Dir. Psychol. Sci. **8**(2), 65–68 (1999)

6. Chohra, A., Chohra, A., Madani, K.: Affect in complex decision-making systems: from psychology to computer science perspectives. In: Iliadis, L., Maglogiannis, I., Papadopoulos, H., Sioutas, S., Makris, C. (eds.) AIAI 2014. IAICT, vol. 437, pp. 174–183. Springer, Heidelberg (2014). https://doi.org/10.1007/978-3-662-44722-2_19

7. Dang, Y., Li, Y., Wu, J.: Quantitative study of affection generated by knowledge exchange in team. J. Inf. Knowl. Manag. **12**(04), 1350036 (2013)

8. Elfenbein, H.A.: Emotion in organizations: a review and theoretical integration. Ssrn Electron. J. **1**(1), 371–457 (2007)

9. Fredrickson, B.L.: What good are positive emotions? Rev. Gen. Psychol. J. Div. Am. Psychol. Assoc. **2**(3), 300 (1998)

10. Fredrickson, B.L.: The role of positive emotions in positive psychology: the broaden-and-build theory of positive emotions. Philos. Trans. R. Soc. Lond. **56**(3), 218–226 (2001)

11. Smith, L.D.: Behaviorism and logical positivism: a reassessment of the alliance. Contemp. Sociol. **25**(3), 467–468 (1986)

12. Hochschild, A.R.: Emotion work, feeling rules, and social structure. Am. J. Sociol. **85**(3), 551–575 (1979)

13. Isen, A.M.: An influence of positive affect on decision making in complex situations: theoretical issues with practical implications. J. Consum. Psychol. **11**(2), 75–85 (2001)

14. James, K., Brodersen, M., Eisenberg, J.: Workplace affect and workplace creativity: a review and preliminary model. Hum. Perform. **17**(2), 169–194 (2004)

15. Junior, N.N.T., Vidotti, A.F., Alaranta, M., Fulk, H.K.: The influence of positive emotions on knowledge sharing. In: Americas Conference on Information Systems (2017)

16. Kelly, J.R., Barsade, S.G.: Mood and emotions in small groups and work teams. Organ. Behav. Hum. Decis. Process. **86**(1), 99–130 (2001)

17. Krylova, K., Vera, D., Crossan, M.: Knowledge transfer in knowledge-intensive organizations: the crucial role of improvisation in transferring and protecting knowledge. J. Knowl. Manag. **20**(5), 1045–1064 (2016)

18. Loewenstein, G., Lerner, J.S.: The role of affect in decision making, pp. 619–642 (2003)

19. Maitlis, S., Ozcelik, H.: Toxic decision processes: a study of emotion and organizational decision making. Organ. Sci. **15**(4), 375–393 (2004)

20. Railsback, S.F., Lytinen, S.L., Jackson, S.K.: Agent-based simulation platforms: review and development recommendations. Simul. Trans. Soc. Model. Simul. Int. **82**(9), 609–623 (2006)

21. Shang, J., Fu, Q., Zoltan, D., Shao, C., Fu, X.: Negative affect reduces performance in implicit sequence learning. PLoS ONE **8**(1), e54693 (2013)

22. Steele-Johnson, D., Beauregard, R.S., Hoover, P.B., Schmidt, A.M.: Goal orientation and task demand effects on motivation, affect, and performance. J. Appl. Psychol. **85**(5), 724–738 (2000)

23. Stephens, J.P., Carmeli, A.: The positive effect of expressing negative emotions on knowledge creation capability and performance of project teams. Int. J. Project Manage. **34**(5), 862–873 (2016)

24. Van Knippenberg, D., De Bode, H.J.M.K., Van Ginkel, W.P.: The interactive effects of mood and trait negative affect in group decision making. Organ. Sci. **21**(3), 731–744 (2010)

25. Zambonelli, F., Jennings, N.R., Wooldridge, M.: Organizational abstractions for the analysis and design of multi-agent system. In: International Workshop, Aose 2000 on Agent-Oriented Software Engineering, pp. 235–251 (2001)

Evolutionary Analysis of Developer Collaboration Network in Cloud Foundry OSS Community

Pengchen Zhang, Peng Liu$^{(\boxtimes)}$, and Nianxin Wang

School of Economics and Management,
Jiangsu University of Science and Technology, Zhenjiang 212003, China
18261936506@163.com, liupeng19821017@126.com,
wangnianxin@163.com

Abstract. The collaborative pattern of developers in OSS (Open Source Software) communities is a research-hotspot in the academic circle. However, the existing researches mainly concern the static features of the communication network of community members, and few studies involve the structural evolution of developer collaboration network in OSS communities. This paper constructs the developer collaboration network of the Cloud Foundry OSS community by coding-collaboration relationships, and then analyses the structure and evolution of the constructed network. The results show that a modular pattern centering on a few developers gradually emerges in the developer collaboration network after an evolutionary process of three stages. Core developers have completed a large proportion of the development work and played a coordinating role in development activities, while the periphery developers submit code to specific sub-projects according to their technical background, which complements the core developers' work. Furthermore, the modules of the developer collaboration network are intrinsically related to the sub-projects and continuously contribute code for the corresponding subprojects during the evolutionary process. These results may deepen our understandings of the collaborative pattern of OSS communities, and also have some reference value for the studies of open collaborative innovation in large scale crowds.

Keywords: Cloud Foundry · Developer collaboration network · OSS community · Structural and evolution

1 Introduction

In recent years, with the maturity and application of web2.0/3.0 technologies, open source software (OSS) has flourished and has been widely used in many fields such as artificial intelligence and big data. Compared with commercial software, OSS shows great autonomy in the development mode. Specifically, in the absence of central

---

Supported by the National Natural Science Foundation of China under Grant No. 71871108 and the Social Science Foundation of Jiangsu Universities in China (Grant No. 2017SJB1092).

J. Chen et al. (Eds.): KSS 2019, CCIS 1103, pp. 87–105, 2019.
https://doi.org/10.1007/978-981-15-1209-4_7

control, numerous community members discover and solve complex problems during the development process through spontaneous collaborations between each other. The collaborative pattern of OSS communities thus plays an important role in the success of OSS development. Correspondingly, the collaborative pattern of OSS communities has attracted considerable attentions in various disciplines, such as computer science [1], management science [2], information science [3] and so on.

For the studies of the collaborative pattern of OSS communities, a pioneering work from Raymond claimed that the development of most commercial software adopts the "cathedral" mode, while the development of OSS represented by Linux utilizes "bazaar" mode [4]. This conclusion has presented a new view of the development mode of OSS. However, the development of OSS, especially the development of large OSS, is a knowledge-intensive project and the interaction between community members has inherent self-governance and complexity [5, 6]. The "bazaar" thus is not enough to uncover the collaborative pattern of OSS communities, which is essentially a knowledge-based adaptive system [2, 7]. Accordingly, scholars conduct a lot of investigations from the perspective of complex networks which can be roughly divided into two categories.

In the first category, the studies mainly concern the structural characteristics of the collaboration network among community members at the macro level [8–14]. For example, Bird et al. constructed the developer collaboration network by using the email communication data of Apache server project, and the results show that the degree distribution of the network has obvious scale-free characteristics [11]. By analyzing the log files and email communication data from the 15 OSS communities on Sourceforge, Singh found the corresponding collaboration networks exhibit different small-world characteristics which are also related to software success [12]. Besides, some scholars have found that the collaboration network of the OSS community has a "core-periphery" structure [13, 14]. Specifically, the network is composed of two kinds of nodes, namely "core" and "periphery", and the core nodes interact closely while periphery nodes hardly interact with each other.

Parallel to these macro-level studies, researches of the second category are conducted to explore the micro characteristics of the two parts of a collaborative relationship [15–18]. For example, Hu et al. analyzed the correlation between individual attributes (e.g. mutual familiarity) and the formation of collaborative relationships between OSS developers [15]. Through investigating eighteen OSS developer networks, Joblin et al. observed a shift from the hierarchical structure to the hybrid structure in the networks (i.e., hierarchy only appears in the relationships of core developers, but the relationships of periphery developers do not have such feature). This result indicates a developer's location in the network influences the establishment of new relationships [16].

The above studies have deepened our understandings of the collaborative pattern in OSS communities to a great extent. However, more efforts are needed to elaborately explore the collaborative pattern of developers in OSS communities, especially in large OSS communities. For one thing, the email communication data are used in the existing work to construct the developer collaboration networks, which may affect the accuracy of results because the information of developers is mixed with that of ordinary users in these data. For another, most of the related studies pay more attention to the static features of the collaboration network. The evolutionary characteristics of the developer collaboration network should not be ignored. Especially, whether the structural properties (e.g., scale-free, small-world, core-periphery) can coexist in the network? How do the structural properties emerge through the network endogenous evolution? What's more, what collaborative behaviors are associated the network evolution? The answers of these questions may further reveal the collaborative pattern and its formation process in OSS communities.

Accordingly, taking the Cloud Foundry community as an example, this paper constructs the developer collaboration network of the Cloud Foundry OSS community by coding-collaboration relationships and then analyses the structure and evolution of the constructed network. We attempt to provide some new thinking on the analysis of OSS communities from the perspective of complex networks. Besides, the development of OSS is a specific form of open innovation and the related studies focus on the fields such as scientific research and patent development, the results of this paper are also complementary to the researches of open innovation.

2   Data and Methods

2.1   Data Collection

Cloud Foundry (CF for short), which is one of the earliest projects in cloud applications and cloud services, is an open-source cloud platform initiated by VMware and managed by the Cloud Foundry Foundation. By January 2019, the project consists of 41 subprojects (e.g. CLI (Official Command Line Client), UAA (User Account & Authentication Server) and Routing). All the code of these sub-projects is committed and modified using git (an open-source distributed version control system), which have recorded each developer's commit records of the project at different periods.

Therefore, git was used in this paper to collect commit records of all subprojects. We have collected 586727 commit records from August 2010 (the initial stage of the project) to January 2019, each commit record contains the author's email, commit time, code modification status and involved files (as shown in Fig. 1).

```
zyjiaobj@cn.ibm.com##Fri Dec 7 11:39:16 2018 +0800##zyjiaobj
2 files changed, 4 insertions(+), 4 deletions(-)
app-autoscaler-release/templates/app-autoscaler-deployment-fewer.yml
app-autoscaler-release/templates/app-autoscaler-deployment.yml
```

Fig. 1. Illustration of one piece of commit records

2.2   Method

Network Construction Method. In this paper, we construct the developer collaboration network through the following steps.

Firstly, the developers are distinguished from one another by email address in each commit record. In other words, if two records are committed through the same email address, we consider the code of the two records is written by the same developer. Correspondingly, the email addresses are denoted as nodes in the constructed network.

Secondly, the coding-collaboration between two different developers is denoted as an edge in the network. Because each subproject of CF is released by versions and the new version is an improvement on the old version, each version can be treated as an independent knowledge product produced by the developers who have contributed code for such version. Thus, two developers are considered to have a collaborative relationship only if they commit code in the same file of the same subproject version. The specific extraction process of the collaborative relationship between two developers is shown in Fig. 2.

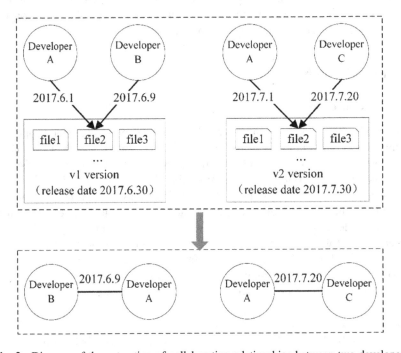

Fig. 2. Diagram of the extraction of collaborative relationships between two developers

In Fig. 2, suppose the release date of the v1 version of a subproject is 2017.6.30, and developer A and B commit code to file2 before such date. Subsequently, after the v1 version release, developer A and C commit code to file2 before the release date of the v2 version. Accordingly, developer B and C respectively create a collaborative relationship with A, while there is no relationship between B and C, the timestamp of the two relationships are respectively set as the commit time of B and C.

Finally, based on the extracted relationships, we construct cumulative collaboration networks of developers at six-month intervals (i.e., the cumulative collaboration networks under seventeen time-windows), such as the collaboration network during 2010.8–2011.1, the collaboration network 2010.8–2011.7, and so on.

Network Analysis Method. We in this paper mainly adopt social network analysis to investigate the constructed collaboration network. Network size is measured by the number of nodes, degree represents the number of neighbors of each node, and the giant component is the largest connected-subgraph in the network.

Besides these above basic indexes, the clustering coefficient, the average shortest-path-length, and the modularity are used to detect the structural characteristics of the collaboration network. Specifically, the clustering coefficient is the amount ratio between triangles and triples in the network while the average shortest-path-length represents the average number of steps along the shortest paths for all possible pairs of nodes. Modularity is used to measure the strength of the division of a network into modules (also called communities). In this paper, the *Louvain* algorithm [19] is used to calculate the modularity of the network.

3   Analysis Results

3.1   Structural and Evolution Characteristics of the Overall Network

Table 1 lists the statistical characteristics of the constructed network where the seventeen time-windows are denoted from *T0* to *T16*. As time-window goes on, the consistent increase of network size indicates the group of developers continuously expands. The average degree maintains around 4.0 after a sharp growth, revealing that the number of partners of each developer tends to be a relatively stable state. Although the entire network is always segregated (the number of connected-subgraphs is much larger than 1 in all time-windows), we can observe a large-scale giant component gradually emerges. Figure 3 intuitively shows the topologies of the collaboration network in different time windows. We can see that the giant component (i.e., the colored subgraph) continuously expands and its structure also changes significantly.

Table 1. Structural characteristics of the overall network

Time-window	Network size	Average degree	Number of connected-subgraphs	Size of giant component	Size of second largest connected-subgraph
T0	4	1	2	3	1
T1	10	0.8	6	4	2
T2	25	0.88	15	8	3
T3	67	2.06	27	32	7
T4	112	2.63	37	56	17
T5	621	4.19	142	448	5
T6	909	4.45	216	650	7
T7	1115	4.15	291	773	9

(*continued*)

Table 1. (*continued*)

Time-window	Network size	Average degree	Number of connected-subgraphs	Size of giant component	Size of second largest connected-subgraph
T8	1361	4.17	362	929	14
T9	1626	4.00	445	1108	9
T10	1818	3.96	504	1226	11
T11	2033	4.00	577	1355	9
T12	2207	4.02	635	1466	19
T13	2376	4.10	679	1586	22
T14	2552	4.15	741	1704	13
T15	2693	4.21	786	1808	13
T16	2798	4.23	829	1872	16

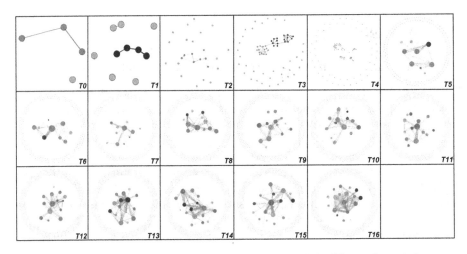

Fig. 3. Topologies of the developer collaboration network in different time windows

To further examine the structural characteristics of the collaboration network, Figs. 4, 5 and 6 respectively draw the degree distribution, the edge-density and the statistics of commit records of the *T16* network.

In the log-log coordinates, Fig. 4 plots the proportion of nodes ($P(k)$) with different degrees (k) in the network. In Fig. 4, with the increase of degree, the proportion of the corresponding nodes (the black circle) declines quickly, and the distribution generally conforms to the power-law distribution with the exponent of −2.83 (the red line). This indicates that a few developers have the majority of collaboration relationships, while most developers have relatively few partners.

Arranging the nodes in descending order of degree (k), Fig. 5 plots the edge-density of the giant component and the isolated nodes (i.e., the nodes outside the giant component) respectively, where the black dash line identifies the giant component and isolated nodes (i.e., the lower-left area of the line is the giant component while the

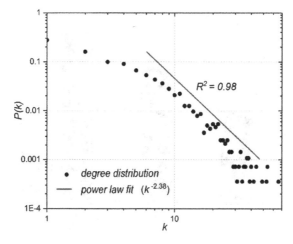

Fig. 4. Degree distribution of developer collaboration network (*T16*) (Color figure online)

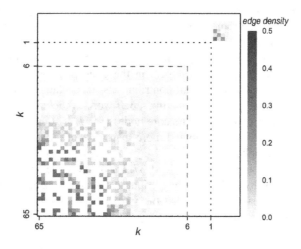

Fig. 5. Edge-density of the developer collaboration network (*T16*) (Color figure online)

right-upper area is the isolated nodes). In the giant component, the edge-density gradually decreases with a downward node degree, which can be basically divided into two regions in the k-k parameter plane (marked by the red dash line). The edge-density of the first region ($k \geq 6$) is significantly higher than that of the second region ($k < 6$), revealing a core-periphery structure in the giant component. Meanwhile, compared with the nodes in the giant component, the commit behavior of isolated nodes has occasional features owing to the extremely sparse edges between themselves.

Figure 6 counts the commit records of three types of nodes observed in Fig. 5 (i.e., the core nodes, the periphery nodes, and the isolated nodes), where the numbers in the horizontal axis have marked the number of different kinds of nodes and their

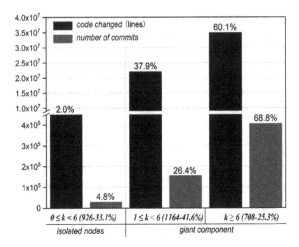

Fig. 6. Code changed and the number of commits of three types of nodes (*T16*)

percentage in the whole network. For example, the number of $k \geq 6$ nodes is 708, accounting for 25.3% of the network size, and more than sixty percent of the code contribution and commits come from these nodes. From Fig. 6, it can be seen that most development work (98% of the code contribution and 95.2% of commits in the whole project) is accomplished by the developers in the giant component. Although these isolated nodes do not make significant contributions, they play a positive role in the implementation of the CF project by sharing a part work of the developers in the giant component. In the giant component, the core nodes (i.e., $k \geq 6$ nodes) accounting for 1/4 of the network size have contributed more than 60% of the development work in the CF project. The periphery nodes (i.e., $1 \leq k < 6$ nodes), whose size is the largest in the three types of nodes, have provided 37.9% of the code and 26.4% commits. Combing with the results in Fig. 5 (the periphery nodes inevitably interact with the core nodes), we can basically assert that the work of these periphery nodes is complementary to that of the core nodes.

3.2 Structure and Evolution Characteristics of the Giant Component

As can be seen from the results in Sect. 3.1, the giant component plays a crucial role in the development activities of the CF project, and its structure shows significant variations under different time-windows. Hence, this section further explores the structural evolution of the giant component from the macroscopic, mesoscopic and microscopic levels.

Macro Level Analysis. Figure 7 shows the evolution of the clustering coefficient (C), the average shortest-path-length (L) and the modularity (Q) of the giant component in the collaboration network under varying time-window, where C and L are the ratios of the actual values to the counterparts of the random networks with the same scale. Through comparing the changes in the above three indexes, the evolution of the giant component can be roughly divided into three stages.

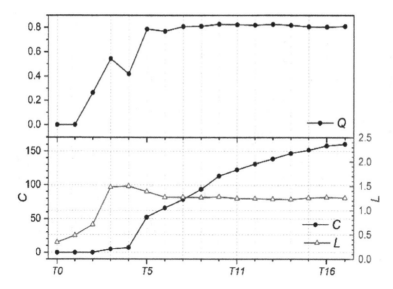

Fig. 7. Evolution of C, L and Q in the giant component

The first stage appears from $T0$ to $T2$, where the values of modularity, clustering coefficient and average shortest-path-length of the giant component are all small ($C \approx 0$, $L < 1.0$, $Q < 0.4$). This implies the edges in the network are sparse so that there is not a clear modular structure. The giant component thus exhibits a state of "loose connection" (e.g., the giant component in the first graph of Fig. 3).

Subsequently, when the time-window moves from $T3$ to $T5$, both the modularity and the clustering coefficient show a rapid growth ($Q \rightarrow 0.8$, $C \rightarrow 50$), the average shortest-path-length remains a high level ($L \approx 1.5$). The giant component here evolves into the second stage, in which a highly modular and "chain-like" structure appears (e.g., the state that the modules in the giant component are connected one by another in the second graph of Fig. 3).

The third stage appears since T6. In this stage, the modularity remains a high level ($Q \approx 0.8$) and the average shortest-path-length has declined ($L \approx 1.3$), indicating modules are interlinked in the giant component. Combining with the continuous increase of the clustering coefficient, it can be asserted that the giant component at this time is a modular small-world network (e.g., the giant component in the last graph of Fig. 3).

Meso Level Analysis. By examining the node-overlaps of the modules in two adjacent time-windows, Fig. 8 depicts the evolutionary process of modules in the giant component, where the black line-segments denote the modules of different time-windows; the length of each line-segment denotes the relative size of the corresponding module in each time-window; the colored flows denote the relationship between two modules which are identified by the node-overlaps. For example, #1 is the largest module in the giant component of T2 and gradually evolves to the #1 module of T3 by merging the #0 module of T2.

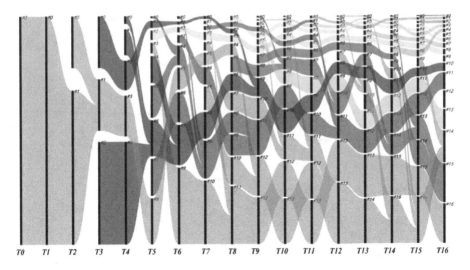

Fig. 8. Evolution of modules in the giant component

In Fig. 8, by observing the relationships between modules, two varying modes can be identified during the module evolution. On one hand, new modules continuously sprout from the giant component (e.g., *#0* and *#2* are the new modules in *T3*). On the other hand, the local modules achieve size expansion mainly through merging small scale modules. For example, module *#4* and #6 in *T5* merge into the second-largest module *#7* in *T6*. Under the combined effects of such varying modes, the giant component maintains a clear modular structure when its size expands.

Based on Figs. 8, 9 illustrates the relations between module evolution and subproject development, where the rectangles represent the modules, the text of each rectangle shows the module number corresponding to that in Fig. 8, the serial numbers of top-4 subprojects in commit amount, the number behind each time-window is the proportion of the number of subprojects that the giant component involved to that of all subprojects in the corresponding time-window. For example, in the time-window *T4*, the top-4 subprojects in commit amount of module *#4* are *16 (cf-release)*, *20 (cloud-controller-ng)*, *10 (bosh)* and *27 (garden)*. T4(1.0) means the giant component has been involved in the development of all subprojects at the moment.

In Fig. 9, all the numbers behind each time-window equal 1, revealing that the developers of the giant component have participated in the development work of all the subprojects. The modules in each time-window always submit code for a few specific subprojects. For example, in time-window *T15*, module *#10* mainly concerns the subproject of *cf-release* while module *#13* makes a lot of code contributions to the subproject of *cli*. Besides, the evolution and merger of modules are also related to the subprojects.

During the module evolution, there is no obvious change in the subprojects concerned by each module and these subprojects often have more functional correlations. Take module *#0* in *T11* as an example, when *#0* gradually evolves into module *#1* in *T16*, the code submitted often touches on subproject *25 (fissile)*, *20 (cloud-*

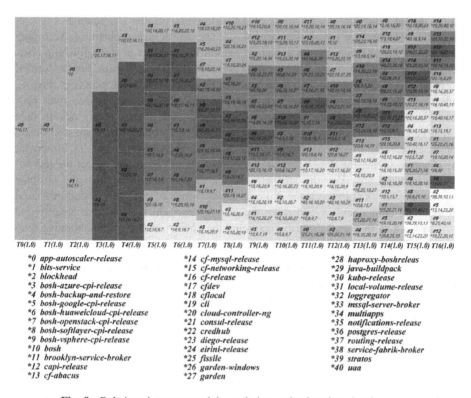

*0	app-autoscaler-release	*14	cf-mysql-release	*28	haproxy-boshreleas
*1	bits-service	*15	cf-networking-release	*29	java-buildpack
*2	blockhead	*16	cf-release	*30	kubo-release
*3	bosh-azure-cpi-release	*17	cfdev	*31	local-volume-release
*4	bosh-backup-and-restore	*18	cflocal	*32	loggregator
*5	bosh-google-cpi-release	*19	cli	*33	mssql-server-broker
*6	bosh-huaweicloud-cpi-release	*20	cloud-controller-ng	*34	multiapps
*7	bosh-openstack-cpi-release	*21	consul-release	*35	notifications-release
*8	bosh-softlayer-cpi-release	*22	credhub	*36	postgres-release
*9	bosh-vsphere-cpi-release	*23	diego-release	*37	routing-release
*10	bosh	*24	eirini-release	*38	service-fabrik-broker
*11	brooklyn-service-broker	*25	fissile	*39	stratos
*12	capi-release	*26	garden-windows	*40	uaa
*13	cf-abacus	*27	garden		

Fig. 9. Relations between module evolution and subproject development

controller-ng), *21 (*consul-release*) and *16 (*cf-release*), and the functions of these subprojects are described as follows. The *fissile* can turn a BOSH release (i.e., an application based on CF) into Docker files. The *cloud-controller-ng* provides an API (Application Programming Interface) to create and manage apps, services, user roles, and more. The *consul-release* is used to provide service discovery, key-value configuration, and distributed locks within cloud infrastructure environments. The *cf-release* contains a canonical BOSH deployment manifest for deploying the CF application runtime. These four sub-projects are all oriented to CF applications, involving the development management, service management, transaction processing, and runtime environment deployment.

For the merger of modules, the merger of two modules is always characterized by the overlaps of the subprojects they mainly maintained. For example, the module #11 in T12 is the product of a merger of the module #1 and #10 in T11 (as shown in Fig. 8), both of which mainly contribute code to the subproject *10, *16 and *7. After the merger, most of commits in #11 are also concentrated on these subprojects.

To sum up, the modules of the developer collaboration network are intrinsically related to the sub-projects and continuously make source-code contributions for specific subprojects. Meanwhile, the giant component achieves the size expansion and modular structure through the emergence of new modules and the merger of local modules.

Micro Level Analysis. As the basis of network evolution, the characteristics of collaborative relationships between developers are also worth further pursuing. According to the previous analysis, the relationships exhibit two aspects of distinguishing characteristics, namely structural characteristics and attribute characteristics. For the former, we have observed a non-uniform distribution of links between developers (e.g., Fig. 4), indicating the two parts of a relationship may have differences between their ego-network structures. For the attribute characteristics, the source code each module committed often concentrates on a few specific subprojects and there are also differences between the involved subprojects of different modules (e.g., Fig. 9). Moreover, the subprojects that the developers are concerned about essentially reflect the technological backgrounds (i.e., the attribute) of the developers. From this, the attribute differences may exist in the inner and inter relationships of modules. Therefore, based on the giant component in *T16* (here the giant component is a modular small-world network), we in this section further examine the characteristics of collaborative relationships from the above two aspects.

Structural Characteristics of Collaboration Relationships. Referring to the work of Guimerà et al. [20], this paper proposed a method to investigate the characteristics of collaboration relationships in the giant component, as shown in (1) and (2) respectively.

In (1), $l_{i,km}$ represents the number of links between k-degree node i and other nodes in the module m, $\bar{l}m$ and σm are the mean value and standard deviation of the inner-module degree of all nodes in the module m, n represents the number of k-degree nodes in module m. Accordingly, the parameter z examines the link distribution of k-degree nodes in the module. A higher value of z means k-degree nodes connect better with the other nodes in module m, otherwise, there are few links between k-degree nodes and other nodes when z value is small.

$$z = \frac{1}{n} \times \sum_{i=1}^{n} \frac{li, km - \bar{l}m}{\sigma m} \tag{1}$$

The p value in (2) mainly measures the role of k-degree nodes in inter-module links between module m and other modules, where lio, km represents the number of inter-module links involved k-degree nodes i, and n is the number of k-degree nodes in module m.

$$p = \frac{1}{n} \times \sum_{i=1}^{n} \frac{lio, km}{k} \tag{2}$$

Figure 10 plots the p and z value of k-degree nodes in each module of the *T16* giant component, where the horizontal axis is the node degree (k) in each module arranged in ascending order of value, and the text above the figure contains the number and size of the corresponding module.

Form Fig. 10, in each module except module #0, the z-value mainly presents an upward trend with the degree increase, and the z-value of $k \geq 6$ nodes is not less than 0. This indicates the core developers (i.e., $k \geq 6$ nodes in Fig. 5) disperse in different

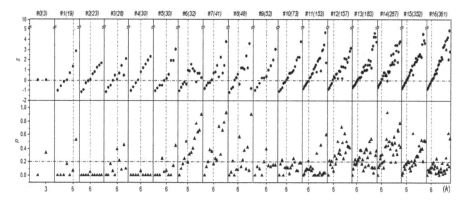

Fig. 10. p and z value of k-degree nodes in each module of the giant component ($T16$)

modules and connect better with the other developers in the corresponding module. In other words, the core developers play a hub role in each module and the inner-module relationships often form centering on their hubs. For the module *#0*, its size is so small that there are no $k \geq 6$ nodes, and the inner-module relationships hardly show any structural characteristics.

With the increase of degree, the changes of p-value do not have any significant regularities in each module. However, by analyzing the p-value of the nodes with different degrees, it can be found that the formation of inter-module links often involves $k \geq 6$ nodes. For example, in module *#12*, the p-value of $k \geq 6$ nodes are higher than 0.2 where the maximum value can reach about 0.7, while the maximum p-value of $k < 6$ nodes is close to 0.2. Although there are no $k \geq 6$ nodes in module *#0*, the nodes with the highest degree (i.e., $k = 3$) also have the maximum p-value. These results indicate the formation of inter-module relationships often requires the participation of core developers (i.e., $k \geq 6$ nodes).

Attribute Characteristics of Collaboration Relationship. To describe the technical backgrounds of developers, we treat the number of commits on different subprojects of each developer as a vector, and then calculate the average attribute-similarity of the nodes with inner-module (or inter-module) links through (3).

In (3), the average attribute-similarity is denoted as S. V_i and V_j represent technical-background vectors of developer i and j, and r represents the serial number of sub-projects. Correspondingly, $V_{r,i}$ means the number of commits on subproject r of developer i. E is the number of inner-module (or inter-module) links.

$$s = \frac{1}{E} \sum_{i \neq j} \frac{\sum\limits_{r=0}^{40} Vr,i \times Vr,j}{\sqrt{\sum\limits_{r=0}^{40} (Vr,i)^2} \times \sqrt{\sum\limits_{r=0}^{40} (Vr,j)^2}} \tag{3}$$

Figure 11 depicts the average attribute-similarity of the nodes with inner-module (inter-module) links in the giant component of *T16*. In Fig. 11, the colored blocks on the counter diagonal represent the *S*-value of the nodes with inner-module links while the other colored blocks represent the *S*-value of the nodes with inter-module links, and the color depth is positively correlated with *S*-value. A block with black slashes means there are no inter-module links between the two corresponding modules.

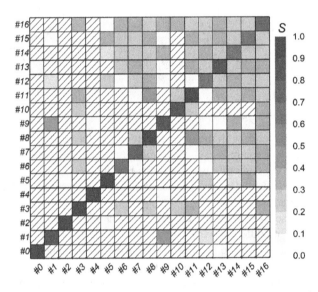

Fig. 11. Average attribute-similarity (*S*) of nodes with inner-module (inter-module) links (*T16*)

It can be observed from Fig. 11 that the color of each block on the counter diagonal is much darker than that of the blocks in corresponding rows (or columns), revealing the nodes with inner-module links often have a higher value of *S* compared with the counterparts with inter-module links. For example, both in module *#4* and *#9*, the *S*-value of the nodes with inner-module links is closed to 0.9. But the *S*-value of the nodes who have created the inter-module links between module *#4* and *#9* is no more than 0.5. This result implies the developers from the same module often have similar technical backgrounds, while there are obvious attribute-differences between the two parts of an inter-module relationship.

From the analysis results of Figs. 10 and 11, we can deduce that the core developers (i.e., the $k \geq 6$ nodes) have played two roles in the process of CF project development. For one thing, the core developers work closely with the others of the same modules (i.e., the hub role within the modules); for another, the core developers collaborate with each other to some extent (i.e., the brokerage role between the modules). The activities of the core developers thus have a coordinating effect on the whole development work, which is essential to the success of CF OSS. Meanwhile, in the giant component, the developers with similar technological backgrounds aggregate to form the different modules; correspondingly, two developers from different modules are different in the technological backgrounds. Hence, the collaborative behavior of the developers displays a combining characteristic of homophily and heterophily.

4 Discussion and Conclusion

4.1 Discussion

Overall Picture of Structural and Evolutionary Characteristics. From the previous analysis, we can obtain the structural and evolutionary characteristics of developer collaboration network in the CF OSS community. The key findings are summarized as follows.

Firstly, the collaboration network explosively expands after a short period of the CF infancy. Meanwhile, the size of the giant component continuously increases that does not exist in the other connected sub-graphs.

Secondly, for the giant component, three stages can be identified during the evolutionary process at the macro level, namely "loosely connected" state, "chain like" structure and "small world" network. When the giant component evolves into a "small world" network, it also exhibits more topological structures. In general, the degree distribution has significant power-law features, implying a hierarchical structure where the highly-connected developers (i.e., high degree nodes) form the core-level and the loosely-connected developers (i.e., low degree nodes) is the periphery-level. In addition, an obvious modular-structure can be identified in the giant component. Especially, the module evolution (i.e., emergence, growth and merge) is inherently correlated with the subprojects. In other words, each module continuously contributes code to specific subprojects and the merge of two modules often takes the overlaps of subprojects they maintained as a prerequisite.

Finally, the collaboration relationships show non-trivial characteristics both in structure and attribute. Specifically, high-degree nodes are spontaneously assigned to different modules and well connected with the other nodes in the same module. Meanwhile, these high-degree nodes also play an important role in the inter-module links. On the other hand, the modules always consist of the nodes with similar attributes (i.e., there are more overlaps between the subprojects which are concerned by the developers of the same module), while the two parts of inter-module collaboration . show certain differences in the attribute.

Here we give a simple explanation for the structural changes of the giant component, especially the emergence of three different states. In the early stage of the CF project, the development work is so fundamental that a small number of developers can finish it by simple collaboration. The giant component thus shows a loosely connected state. With the increase of software functions, the developers focusing on similar functions closely collaborate with each other, which facilitates the formation of modules in the network. When the number of new functions is small, they can be integrated with the existing ones through a few coordination efforts. Hence, at this stage, different modules are connected one by one, and a chain-like structure emerges in the giant component. With the further increase of new functions, more coordination efforts lead to the emergence of inter-connected module structure and the giant component evolves into a modular small-world network.

Comparison with Relevant Work. As mentioned in the section of introduction, although the studies on the evolution of collaboration network among OSS developers is limited, both Xia et al.'s and Joblin et al.'s work are similar with our study, as these two studies focus on the adaptive structure of collaboration network in the development practices. However, there are essential differences between our work and the two contributions.

Joblin et al. collected the commits of 18 OSS projects (e.g., Node.js, GCC, etc.) and constructed the corresponding collaboration network of developers by the semantic relationship between program functions. The size of these networks ranges from 50 to 1000, which is smaller than that of CF developers in our study. For the network evolution, the networks in Joblin et al.' work exhibit an evolutionary process from loosely connected groups to strongly connected groups, which is also observed in our study. Besides, we identify one more structural state between such two states in the collaboration network of CF developers, namely chain-like structure. This difference implies that the developer collaboration network of different OSS scales may have different evolutionary modes. Especially, large-scale OSS (e.g., CF) has much higher level of software complexity which requires a more specialized division of labour, while the developers of small-scale OSS could have more energy in different software functions. Consequently, the developer collaboration network of large-scale OSS has clear modular structures and exhibits an evolutionary process where the modules are continuously interconnected, while that of small-scale OSS often evolves into a single cohesive module.

For the network size and the software scale, our study is much more similar to Xia et al.'s work, which focuses on the evolution of network communities (i.e., the modules in our study) in OpenStack (OS for short) OSS community. It is notable that both the evolutionary modes (i.e., emergence, growth and merge) of network communities (or modules) are consistent. Nevertheless, for the association subprojects, each network community of OS is more focused than that of CF. Besides, we in CF developer collaboration network do not find the merge of two modules which focus on totally different subprojects, but such phenomenon exists in the OS developer collaboration network. The reason for these differences lies in the service of these two OSS. Comparing with the IaaS (i.e., Infrastructure as a Service) of OS, CF provides PaaS (i.e., Platform as a Service) which requires more complex interaction among different services (e.g., virtual sever and operation system). Correspondingly, the developers of CF give more ongoing effort to govern the relationships between the subprojects they focused and other association subprojects. This leads to the fact that the modules of CF community are less focused than that of OS community in the aspect of involved subprojects. On the other hand, because of the infrastructure service of OS, the fundamental changes may induce community merge which focus on different subprojects. But the modules involving different functional subprojects are often hard to merge in the situation of platform services of CF.

4.2   Conclusion

Taking the Cloud Foundry OSS community as an example, this paper constructs the developer collaboration network by coding-collaboration relationships, and then analyzes the structural evolution of the constructed network. The results show that with the network size expansion, a giant component with the "core-periphery" structure gradually emerges in the collaboration network. Furthermore, the "core" developers with the smallest population have accomplished more than half of the development work, and most of the rest development work is shouldered by the "periphery" developers, who have the largest population in the entire network. For the developers outside the giant component (i.e., the isolated nodes), they occasionally contribute code to the project. Thus, during the CF project implementation, the developers in the giant component are the main force, in which the "core" developers act as the skeleton, while the developers outside the giant component participate in the development work as a reserve force. Besides, the giant component has the following structural and evolutionary characteristics.

The giant component evolves from a "loosely associated" state to a "chain-like" structure, and then to a highly modular small-world network. During such an evolutionary process, the modules are intrinsically related to the sub-projects, including the emergence of new modules, the evolution and merge of local modules.

For the modular small-world structure of the giant component, the developers with similar technological backgrounds aggregate to form the different modules; the core developers are not only the hubs in each module but also the brokers between modules, correspondingly, the main part of each module is composed of periphery developers.

Through the above results, we can draw a holistic view of the work mode of the Cloud Foundry OSS community. Specifically, a modular pattern centering on a few developers gradually emerges in the developer collaboration network: Firstly, apart from the code contribution, the core developers coordinate the whole development work. Secondly, the periphery developers gathered based on technological backgrounds and continuously contribute code to specific subprojects. Lastly, the rest developers (i.e., developers who are not in the first two categories) occasionally contribute code, but they play a positive role in the implementation of the whole project by sharing a part work of the core and periphery developers. Therefore, the collaborative pattern at least in the Cloud Foundry community is not the flat and free-wheeling organizational-structure proposed in the existing researches (e.g., ref. [4]), but the tradeoff between complete freedom and strict division of labor. In other words, the developer can freely choose which subproject to participate, according to their technological backgrounds; but the type of work (e.g., the pure code contribution or coordination) is not shifted by the personal preference, which is usually decided by the amount of code-contribution and the status in the network.

The above results may be helpful to further understand the collaborative pattern and the intrinsic governance structure of large OSS communities. Meanwhile, as a specific open innovation activity, these results also have certain reference value for the studies on open innovation of large crowds. However, there is still a gap between our results and general theories because of the single case in this paper. Thus, more cases from different fields (e.g., the Tensorflow project in artificial intelligence and the AngularJS project in web development) will be added to the future analysis to further examine the results observed in this paper. In addition, combing with the characteristics of collaborative relationships, we are to develop multi-agent models to explore the underlying mechanisms of the developer collaboration network evolution in OSS communities.

Acknowledgment. This study is partly supported by the National Natural Science Foundation of China under Grant No. 71871108 and the Social Science Foundation of Jiangsu Universities in China (Grant No. 2017SJB1092).

References

1. He, P., Li, B., Yang, X.H., et al.: Research on developer preferential collaboration in open-source software community. Comput. Sci. **42**(2), 161–166 (2015). (in Chinese)
2. Xia, H.X., Zhang, X., Zhang, X.Z.: Study on collaborative network of OpenStack OSS developers. Syst. Eng. Theory Pract. **37**(5), 1373–1382 (2017)
3. Gencer, M., Oba, B.: Taming of "Openness" in software innovation systems. Int. J. Innov. Digit. Econ. **8**(2), 1–15 (2017)
4. Raymond, E.: The cathedral and the bazaar. Knowledge. Technol. Policy **12**(3), 23–49 (1999)
5. Shah, S.K.: Motivation, governance, and the viability of hybrid forms in open source software development. Manag. Sci. **52**(7), 1000–1014 (2006)
6. De Laat, P.B.: Governance of open source software: state of the art. J. Manag. Gov. **11**(2), 165–177 (2007)
7. Behfar, S.K., Turkina, E., Burger-Helmchen, T.: Knowledge management in OSS communities: relationship between dense and sparse network structures. Int. J. Inf. Manag. **38**(1), 167–174 (2018)
8. Hong, Q., Kim, S., Cheung, S.C., et al.: Understanding a developer social network and its evolution. In: 2011 27th IEEE International Conference on Software Maintenance (ICSM), Williamsburg, pp. 323–332. IEEE Press (2011)
9. Yang, J., Li, H., Liao, H., et al.: Localization of information on communication networks of an open-source online community. Int. J. Mod. Phys. C **28**(07), 1750091 (2017)
10. Ye, P.G., Mao, J.H., Liu, X.F.: Quantitative analysis of open source project in GitHub community based on big data. Electron. Meas. Technol. **40**(8), 84–89 (2017). (in Chinese)
11. Bird, C., Pattison, D., D'Souza, R., et al.: Latent social structure in open source projects. In: Proceedings of the 16th ACM SIGSOFT International Symposium on Foundations of software engineering, Atlanta, pp. 24–35. ACM Press (2008)
12. Singh, P.V.: The small-world effect: The influence of macro-level properties of developer collaboration networks on open-source project success. ACM Trans. Softw. Eng. Methodol. **20**(2), 1–27 (2010)
13. Crowston, K., Shamshurin, I.: Core-periphery communication and the success of free/libre open source software projects. J. Internet Serv. Appl. **8**(1), 10 (2017)
14. Geldenhuys, J.: Finding the core developers. In: 2010 36th EUROMICRO Conference on Software Engineering and Advanced Applications, Lille, pp. 447–450. IEEE Press (2010)
15. Hu, D., Zhao, J.L.: Discovering determinants of project participation in an open source social network. In: 30th International Conference on Information Systems 2009, Phoenix, p. 16. AIS Press (2009)
16. Joblin, M., Apel, S., Mauerer, W.: Evolutionary trends of developer coordination: a network approach. Empir. Softw. Eng. **22**(4), 2050–2094 (2017)
17. Kavaler, D., Filkov, V.: Stochastic actor-oriented modeling for studying homophily and social influence in OSS projects. Empir. Softw. Eng. **22**(1), 1–29 (2016)
18. Wang, W.J., Li, B., He, P.: An analysis of the evolution of developers' role in open-source software community. Complex Syst. Complex. Sci. **12**(1), 1–7 (2015). (in Chinese)

19. Blondel, V.D., Guillaume, J.L., Lambiotte, R., et al.: Fast unfolding of communities in large networks. J. Stat. Mech: Theory Exp. **2008**(10), P10008 (2008)
20. Guimerà, R., Amaral, L.A.N.: Cartography of complex networks: Modules and universal roles. J. Stat. Mech: Theory Exp. **2005**(02), P02001 (2005)

A Model of Minority Influence
in Preferential Norm Formation

Qiang Liu[1], Hong Zheng[1], Weidong Li[2(✉)], Jiamou Liu[2],
Bo Yan[1], and Hongyi Su[1]

[1] Beijing Lab of Intelligent Information Technology, School of CS and Technology,
Beijing Institute of Technology, Beijing, China
hongzheng@bit.edu.cn
[2] School of Computer Science, The University of Auckland, Auckland, New Zealand
wli916@aucklanduni.ac.nz

Abstract. Minority influence takes place when a minority group influences the majority group, which has been widely studied in psychology and sociology. Most existing works are, however, qualitative studies or based on field experiments; large-scale quantitative experiments are hard to conduct due to the lack of a model. Thus, we propose a new agent-based framework, Minority Influence Model, for modelling the interactions among individuals to explore minority influence in the context of norm emergence. What makes the model novel is that it is an integration of three key aspects. First, the model considers multiple player interactions; second, the actions of agents are driven by a combination of endogenous preference and exogenous social surrounding; third, the decision-making strategy for each agent is based on the evolution of gist. With Minority Influence Model, considerable experiments have been performed. The experimental results suggest that the size of the minority group, the extent to which the agents of the minority group prefer an action to the other, and the degree of disagreement between the majority and the minority can be the three main factors of minority influence. We also found some tipping points under some specific scenarios.

Keywords: Minority influence · Social convention · Social norm

1 Introduction

Emerson wrote that "all history is a record of the power of minorities and of minorities of one"[1]. The power of minorities, called *minority influence*, takes place when one or more members of a minority group influence the members of the majority group to accept or conform to the behaviour of the minority group [6]. Understanding minority influence gives us insight into social events and social movements. For example, it can help us to understand the feminist

---

[1] The Later Lectures of Ralph Waldo Emerson, 1843–1871. University of Georgia Press. 2010.

© Springer Nature Singapore Pte Ltd. 2019
J. Chen et al. (Eds.): KSS 2019, CCIS 1103, pp. 106–121, 2019.
https://doi.org/10.1007/978-981-15-1209-4_8

movement in the last century, which was led by a minority group and persuaded the majority. Also, exploring the factor of minority influence may give us ideas of how to promote or prevent the influence of the minority. For instance, for companies planning to launch new products, it may be helpful to analyse how to attract more users and then dominate the market.

There are considerable studies regarding factors affecting minority influence in the human sciences [13,19,21,27]. most of which are qualitative research. It has been claimed that people's actions are influenced by both their attitudes and perceived social pressure [1,14,18,20]. In groups where individuals are able to interact with each other, individuals under real or imagined social pressure may change their behaviours to be consistent with social conventions [19]. At the same time, individuals holding attitude different from the majority can influence the whole group, if they are consistent and self-confident, and individuals by the minority defect the majority group [19]. However, due to the lack of a model that simulates communication among individuals and the decision-making process of each individual, quantitative studies are hard to conduct.

Consider a scenario with three individuals, Alex, Bob and Charlie. Charlie prefers playing football, while Alex and Bob rate playing football and shopping the same. Additionally, they do not want to go out alone, so they expect some others can choose the same activity as they do. Ideally, they choose the preferred activity and let others choose the same action. For instance, Charlie wants to play football and persuade Alex and Bob to join him. If others do not join him, Charlie might rather go shopping with them. In other words, individuals consider both their preferences and the actions of others when they make decisions. In this example, if Charlie, as the minority, can affect the action of others, the minority influence occurs. The question lies in how to model this scenario.

This scenario requires a framework which is able to model the interactions among multiple individuals, and take the preference of each and the influence of the group on each into consideration. To achieve this, we build the Minority Influence Model and then investigate minority influence in the context of norm emergence. Norm emergence is a well-established field of research in the multi-agent system that describes the emergence of coordinated actions among multiple interacting agents. Through engaging in repeated coordination games, agents may be able to carry out social learning, updating individual actions according to self-interest, while eventually arriving at unity in their action. Minority influence can be regarded as the emergence of the social norm consistent with the action of the minority group. For example, after interacting with each other, Alex and Bob can be affected by the minority Charlie, and choose to play football with him, then a social norm, as well as minority influence, occurs. Thus, a model of norm emergence can be used to investigate minority influence. For any model of norm emergence, there are two main ingredients. The first is the way how agents interact, and the second is the way how agents make decisions. In this paper, the Minority Influence Model is built as follows:

To define the way in which agents interact, we use a *multi-player coordination game with preference*. This game involves a number of players each of which having the same 2 actions to choose from, and their payoffs are determined by (1) their own preference over these two actions and (2) the percentage of agents in the game who had the same action as theirs. The underlying motivation for the separation between preference and actions, as well as the concept of social conformity, is backed by social psychological studies [1,19,20].

To define the way in which agents make decisions, we use a *gist-based reinforcement learning* paradigm. The paradigm assumes the following: (1) agents are able to observe the actions from others that they interact with in the past and (2) such past information is not precisely stored, but rather, is abstracted into a form of "gist", general belief regarding the public's opinion. Inspired by the fuzzy trace theory [2], the notion of gist-based Q-learning was recently proposed in the work of Hu et al. [8], where the Q-value of each action can be determined by current perceived gist. It is a major novelty in this paper that this paradigm is used in the context of the multi-player coordination games with preference.

The combination of the two ingredients above has resulted in a new model that enables the investigation of minority influence. The significance of this new model is that it is the first which integrates multiple aspects: (1) multiple player interaction (2) agents actions are driven by a combination of endogenous preference and exogenous social surrounding (3) the evolution of gist in explaining the formation of social norms. To explore minority influence, we conduct a set of experiments to reveal potential correlations between the many factors involved and the outcome of the social norm. The experimental results suggest that minority influence can be correlated with the size of the minority group, the degree to which the minority prefers some certain action, and the degree of disagreement between the majority and the minority. Tipping points can be observed under two scenarios. The first is that the majority has a neutral attitude, and a contrary action to the minority. The second is that the minority holds a contrary preference to the majority.

The paper is organised as follows: In Sect. 2, we review related literature; in Sect. 3, we propose the Minority Influence Model; in Sect. 4, we present a set of experiments based on the proposed model; in Sect. 5, we draw a conclusion and suggest some possible future work.

2   Related Work

Numerous studies in psychology and sociology have analysed minority influence. Myers regarded minority influence as the fact that individuals can influence their groups [19]. Milgram et al. claimed that a group with a larger size could be more influential [16]. Moscovici argued that the minority who are consistent in its attitude could be more influential than a wavering minority [17]. Allen and Levine claimed that when the minority doubt the attitude of the majority consistently, some of the members of the majority can defect to the minority. Once switch to the minority, they can be more influential than initial members of the minority [13]. Furthermore, once defections start, members of the majority can soon follow the defector and switch to the minority [19].

However, most of the studies above are qualitative studies. To conduct quantitative experiments, a model is required. Minority influence can be viewed as the process of the majority conforming to the action of the minority, which in fact establishes a social norm. Thus, models of norm emergence can be used to explore minority influence. Early studies regarding norm emergence introduced the notion of emergent conventions [23]. In 1993, Kittock adopted 90% convergence metric and claimed that a convention emerges in a system if at least 90% of the agents chooses the same action [12], which is widely used in studies of norm emergence as well as this paper. Models of norm emergence consist of two main ingredients. The first is the way that agents interact, and the second is the method that agents make decisions.

In terms of the interaction ingredient, there are extensive research works. In 1997, Shoham and Tennenholtz defined standard game-theoretic notions such as n-k-g stochastic game [24]. n-k-g stochastic game illustrates the repeated process of randomly selecting k agents from n given agents and letting them play a certain base game. This framework has been widely used in the domain of norm emergence and also adopted in this paper.

Coordination game [12] can be used as the base game of n-k-g stochastic game. In a coordination game, all agents can receive the same payoff if they choose the same action, which indicates that the goal of the agents is to achieve uniformity [5,24]. Coordination game ignores the fact that individuals can have different preferences on different actions in real life. To enrich this model, several variants have been introduced, one of which is unbalanced coordination game [5]. Compared with standard coordination game, agents can receive different payoffs when they converge to different actions. It considers the difference between actions, while it is still not suitable for scenarios where individuals have different preferences. Another variant is competitive-coordination game [9], also known as the battle of the sexes [5], which is proposed to simulate the situation where agents tend to build different conventions, although they intend to convention emergence. In other words, agents prefer coordination than dis-coordination, but different agent prefers different coordination. Competitive-coordination game considers the different preference of each agent, but it is not suitable for modelling the situations where more than two agents are involved. One more variant is n-player coordination game [8], which generalises the standard coordination game and can be used to model the interactions among more than two individuals. The payoff each agent receiving is based on the proportion of other agents choosing the same action. For each agent, the more other agents choosing the same action as it, the more payoff it receives. However, n-player coordination game does not consider the different preference of each agent. Therefore, to take both endogenous preference and exogenous social surrounding into consideration, we combine competitive-coordination game and n-player coordination game to establish the interaction ingredient of the Minority Influence Model.

Decision-making strategy is another essential part of norm emergence models. As individuals learn from experience [24], the decision-making strategy essentially represents the "brain" for each individual. In early studies, memory-based strategies such as the Highest Cumulative Reward (HCR) update rule has been used as a decision-making method [24]. Recently, researchers apply reinforcement learning algorithms to decision-making strategies [8]. Reinforcement learning allows an agent to learn a sequence of actions through trial-and-error interactions with a dynamic environment [10]. Q-learning is a model-free reinforcement learning which dynamically evaluates the available actions [26]. It has been shown that Q-learning with ϵ-greedy exploration leads to faster convention emergence compared to other methods [22]. Thus, many works try to extend Q-learning. Hao and Leung extended the idea of joint action learner [4] to norm emergence problem, and proposed a decision-making method that agents learn the q-values of joint actions, i.e., a combination of their own actions and the actions of their neighbours [7]. Hu et al. modified Q-learning to establish gist-based Q-learning [8], based on the fuzzy trace theory [2].

Gist is a psychological concept that refers to a type of vague and high-level mental representation of events [2]. In real life, individuals tend to memorise events by gist, i.e., memorise only general information instead of details. Inspired by this, gist-based Q-learning has been proposed [8]. At each iteration, agents update their Q-values based on the perceived prevalence of actions and choose the action with the higher Q-value. The perceived prevalence is implied from the actions of other agents it interacts with at this iteration. Gist-based Q-learning can reflect the influence of the actions of other individuals on one single agent, which is consistent with our assumption in this paper. Thus, we borrow and modify gist-based Q-learning in the Minority Influence Model, which is introduced next.

3 Minority Influence Model

Imagine a situation, say, that a few individuals are discussing whether to buy a new mobile phone, and there are only two decisions available, to buy or not to buy. Due to different background, personality, and position, people can hold different opinions about this mobile phone. Some people might prefer to buy the phone because they need a new phone or like the manufacturer, while others can prefer not to buy it because they do not like the design of this phone or they are fans of another brand. Additionally, people can not only favour different decisions, but they can prefer different decisions to different extents. Some people may feel it is acceptable to either purchase the phone or not purchase. In contrast, some people extremely want to own the new product, and some individuals can hate the phone.

On the other hand, people tend to follow the action of the majority to avoid punishment and get a benefit [3,15,19,20,28]. In this scenario, a possible consequence can be isolation from friends, when one person decides not to buy the phone, while most of his friends want to buy it. Thus, people need to balance their preferences and the influence of the majority to make the right decision, which can bring more benefit. However, people in the real world cannot know the opinion of each person in society. Instead, they can only get the opinion of

---

Algorithm 1. The Framework of the Minority Influence Model

Input agents N, number of iterations T, number of peers k
initialisation (N)
$t = 0$
while $t < T$ **do**
  for $agent \in N$ **do**
    randomly select k agents from $N \setminus \{agent\}$, set them to P
    $players = \{agent\} \cup P$
    n-player coordination game with preference (players)
  end for
  $t = t + 1$
end while

---

people whom they have met and communicated. Their knowledge of the majority is derived from their observation of the people around them. As a result, their perception of the opinion of the majority can change as they interact with new people or people who change their ideas.

What we are interested in is that under the circumstance discussed above, whether a minority of people with extreme opinions can influence the whole society. In order to explore this problem, we propose a new model, the Minority Influence Model, to simulate how people get influenced by their own preferences and others' opinions. We first introduce the framework of the Minority Influence Model, then explain two mechanisms applied in the model, namely n-player coordination game with preference that simulates how people interact with each other, and decision-making method that mimics how people make decisions in this particular scenario.

3.1 The Framework of the Proposed Model

This framework (Algorithm 1) is an overview of the model which simulates how people get influenced by their own preferences and others' behaviours. Assume that there are n people in the society and they can interact with some others every day. Here we apply repeated game, which means some base game, i.e., n-player coordination game with preference is repeatedly executed. After t days' interaction, we observe whether the minority can influence the majority.

To formalise this process, we use N to denote the set of individuals $a_1, \ldots, a_n$, T to denote the number of iterations, $A = \{x, y\}$ to represent the two actions available to each agent. Firstly, we initialise N. Each agent a_i is given two values α and β, which denote the extent to which the agent prefers action x and y, respectively. Then, for each agent, we randomly select k agents as its peers or neighbours, and let the $k + 1$ agents play the n-player coordination game with preference. We repeat this process for T times. More details about the n-player coordination game with preference are given in the next section. Finally, we observe the actions of each agent.

3.2   N-Player Coordination Game with Preference

N-player coordination game with preference is the core part of the Minority Influence Model. It reflects how a group of people share their own opinions or decision about one topic, influence each other, and get influenced by each other, then evaluate their own opinions or decisions. It formalises the interactions between individuals, considering both the power of the preference of each individual and the opinion of the majority.

For example, let us consider a 4-player coordination game with preference. Assume that these four individuals discuss whether to buy a newly-listed mobile phone. In the beginning, only one person wants to buy this phone, because he is a fan of the company which designed the product. Other three people are fans of another company, so decide not to buy it. For the person who wants to purchase the phone, he evaluates his decision by considering both his own taste and others' opinions. If he purchases the phone, he can be happy because of making a decision he preferred, but others may judge him and make him feel isolated. If he does not buy the phone, he can integrate into the group, but cannot get the product he loves. After consideration, he still wants to purchase the phone, because he is loyal to the company. For other people who do not want buy the phone, their tastes are consistent with the opinion of the majority, so they do not change their decisions.

To formalise this process, we propose n-player coordination game with preference. We use agents to denote individuals, actions to denote the opinion or decision of individuals, and rewards to represent the result of the evaluation on decisions of each individual. Individuals prefer holding the same opinion as others than holding a different idea, so we say that one individual prefers the convention λ_x if he prefers the action x.

Let N be the set of all individuals, $A = \{x, y\}$ be the set of decisions available. N_x be the set of agents prefer convention λ_x, and N_y be the set of agents prefer convention λ_y, such that N_x and N_y are disjoint sets, and $N_x \cup N_y = N$.

Definition 1. *An n-player coordination game with preference is a 4-tuple*

$$\langle N, A, (p_i), (r_i) \rangle,$$

where $N = \{1, 2, \cdots, n\}$ is the set of n agents; $A = \{x, y\}$ is the set of actions that agents can perform; p_i is the convention that an agent i prefers, i.e., $p_i = x$ iff $i \in N_x$; (r_i) is the reward function which is defined as follows:

$$r_i = \begin{cases} \alpha \times \dfrac{\sum_{j=1}^n I(a_i, a_j) - 1}{n - 1}, & \text{if } a_i = p_i, \\[4mm] \beta \times \dfrac{\sum_{j=1}^n I(a_i, a_j) - 1}{n - 1}, & \text{otherwise}, \end{cases} \qquad (1)$$

where α and β are the preference values of the preferred action and the other action of the agent i respectively, $\alpha > \beta > 0$, and $I : A \times A \to \{0, 1\}$ is a function such that:

$$I(a_i, a_j) = \begin{cases} 1 & if\ a_i = a_j, \\ 0 & otherwise. \end{cases} \tag{2}$$

For each agent of this coordination game, the preference values α and β are the maximum rewards that it can receive by performing the preferred action and the ill-favoured action, respectively. They denote to what extent an agent prefers some action or convention. The actual reward it receives is calculated by the corresponding preference value multiplies the proportion of its neighbours who are performing the same action as it. The reward depends on both the action it chooses and the decisions of other agents involved in this game, which is supported by psychological studies [1,20]. If an agent decides to choose action x, the more other agents choose action x, the higher reward this agent receives. Similarly, if an agent can predict that most of its peers will choose action x, it can then compute the possible rewards of performing each action, then choose the action with a higher reward.

It is not necessarily true that the agent performing the preferred action, which results in a higher preference value, leads to a higher reward. For example, consider a 4-player coordination game with preference. Let agent a_1 prefers action x with a preference value 8, and a preference value 2 for another action y. Let other agents a_2, a_3 and a_4 perform action y. If agent a_1 performs action x, the reward it receives is $8 \times (0/3) = 0$, and if it chooses action y, the reward it receives is $2 \times (3/3) = 2$. Similarly, it is also not necessarily true that a convention where an agent chooses the action choosing by the majority of the group can bring a higher reward than a non-convention.

N-player coordination game with preference models the process of people influencing each other during the interaction. At the end of each game, people can receive some feedback or evaluation of their decisions (the rewards). People tend to learn from each game to make a better decision in the future. Next, we introduce the mechanism used by agents to make better decisions.

3.3 Decision-Making Under Gist-Based Q-Learning

As we discussed before, people's decisions can be influenced by their own preferences and the opinions of other people. As gist-based Q-learning [8] illustrates how people make decisions under the influence of others, we modify this algorithm to build the decision-making mechanism of the proposed model.

The same as previous, each individual has two possible actions x and y. The reward they received is computed by the mechanism introduced before. We assume all people are gist-neutral, which means they are objective so that their perceived prevalence of actions is the actual frequency of actions [8].

Let $A = \{x, y\}$ be the set of actions that available for each agent, following gist-based Q-learning, the Q-value of an action x at time t is calculated as follows:

$$Q_x^t = b_x^t + w_y^t p_y^t \tag{3}$$

where b_x^t is a bias parameter that describes how well to perform action x without considering others' decisions, w_y^t is a weight that reflects the correlation between

---

Algorithm 2. The Modified Gist-based Q-learning

Input the performed action x, the perceived prevalence of the other action p_y, reward r, learning rate η

$Q_x = b_x + w_y p_y$

$b_x = b_x + \eta(r - Q_x)$

$w_y = w_y + \eta(r - Q_x)p_y$

$Q_x = b_x + w_y p_y$

---

the Q-value of action x the perceived prevalence of action y, and p_y^t denotes the proportion of agents choosing action y in the neighbourhood. As we assume that all agents are gist-neutral, p_y^t can be used to represent the perceived prevalence of action y.

Q-values for each action denote expect rewards of performing these actions. Thus, at each iteration, agents perform the action with a higher Q-value. During the coordination game, they observe others' actions, and then update their q-table. The method to update Q-values is introduced as follows.

In gist-based q-learning, instead of updating Q-values directly, agents learn the bias parameter b and the weight w to minimise the deviation between the Q-values of actions and the actual reward the agents receive by choosing these actions. Gradient descent is used here to learn the values of b and w. Assume that an agent chooses action x at this iteration, gets the reward r, and observes the proportion of people choosing the other action p_y. Algorithm 2 shows how we update the values of b and w, and then update the Q-value for the agent.

For instance, an agent chooses the action x, gets a reward 5, and observes 50% of his peers choose the other action y. Set the learning rate η to 0.1, and assume the initial value of b_x and w_y are 4 and 1, respectively. We show the procedures as follows. Firstly, compute the Q-value: $Q_x = b_x + w_y p_x = 4 + 1 \times 0.5 = 4.5$. Secondly, update the value of b_x: $b_x = b_x + \eta(r - Q_x) = 4 + 0.1(5 - 4.5) = 4.05$. Thirdly, update the value of w_y: $w_y = w_y + \eta(r - Q_x)p_y = 1 + 0.1(5 - 4.5) \times 0.5 = 1.025$. Finally, update the Q-value: $Q_x = b_x + w_y p_y = 4.05 + 1.025 \times 0.5 = 4.5625$. After this process, the agent updates its expected reward for action x. At the next iteration, the agent can choose the action with a higher Q-value, get a reward, then update its q-table again.

To sum up, the Minority Influence Model consists of a base game and a decision-making strategy. By repeating the base game, agents can learn from interactions and may switch to another action. The action they choose is based on their own preferences and the actions of others. Based on this framework, in the next section, we conduct experiments to explore minority influence.

Table 1. Parameters for experiments

Parameter	Description	Default value
n	The number of agents	100
T	The number of iterations	1000
E	The number of repeated experiments	100
p	The proportion of the minority agents	0.15
α_0	The preference value of action x for the minority agents	8
β_0	The preference value of action y for the minority agents	2
α_1	The preference value of action x for the majority agents	5
β_1	The preference value of action y for the majority agents	5
k	The number of peers in coordination games	7
η	The learning rate	0.01
ϵ	The explore rate	0.05

4 Experiments

4.1 Experiment Setup

Following Minority Influence Model, first we define all the parameters in Table 1. In each set of experiments, we may only modify several parameters, and others that are not mentioned remain the default value.

In each set of experiments, agents are divided into two groups, the minority, and the majority. Agents of the minority prefer action x than y, and they always choose action x at the beginning. Initially, both the minority and the majority groups achieve Nash equilibrium, where no agent can gain more by changing its strategy [25]. Then individuals of the two groups begin to interact with each other. After a number of iterations, we observe whether the minority can influence the majority. Since the preference of the majority can be consistent, neutral, or contrary compared to the minority, we consider three scenarios, which are introduced in more details next.

Scenario 1: Consistent Preference Scenario. The first scenario assumes that the agents of the majority share the same preference with the minority but act differently at the beginning. We set $(\alpha_0, \beta_0) = (\alpha_1, \beta_1)$, where $\alpha_0 > \beta_0$, the initial action of the minority to x, the initial action of the majority to y.

Scenario 2: Neutral Majority Scenario. Different from the first scenario, the majority may have a neutral attitude towards the actions, i.e., they prefer the two actions at the same level. Thus, we set $(\alpha_1, \beta_1) = (5, 5)$. There are three possible initial states for the majority. The first is that all agents of the majority choose action y. The second is that half the agents choose action x, and others choose y. The third is that all agents choose action x. All three states are under the Nash equilibrium. As the agents of the minority group choose action x, we only consider the first and the second states.

Scenario 3: Contrary Preference Scenario. This scenario assumes all agents of majority have a contrary preference of action compared to the minority group, and the majority will choose their action according to their own preference at the beginning of the iteration. Under this situation, the minority faces higher difficulty to affect the majority. We'd like to explore the case when the minority group has an extreme belief of their preference.

In order to explore the minority influence under different scenarios, we conduct E simulations for each different set of parameters for each scenario. Part of the experiment results is shown and discussed next.

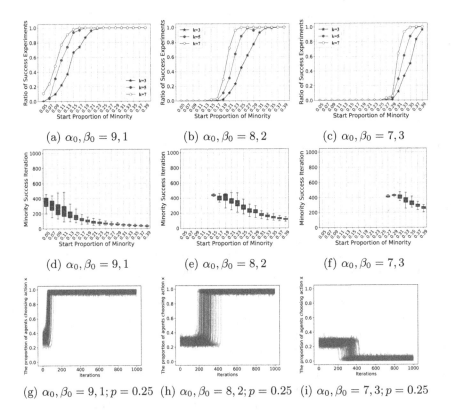

(a) $\alpha_0, \beta_0 = 9, 1$ (b) $\alpha_0, \beta_0 = 8, 2$ (c) $\alpha_0, \beta_0 = 7, 3$

(d) $\alpha_0, \beta_0 = 9, 1$ (e) $\alpha_0, \beta_0 = 8, 2$ (f) $\alpha_0, \beta_0 = 7, 3$

(g) $\alpha_0, \beta_0 = 9, 1; p = 0.25$ (h) $\alpha_0, \beta_0 = 8, 2; p = 0.25$ (i) $\alpha_0, \beta_0 = 7, 3; p = 0.25$

Fig. 1. Scenario 1. Consistent preference scenario

4.2   Scenario 1: Consistent Preference Scenario

The experiment results are displayed in Fig. 1. In Figs. 1a, b and c, the x-axis denotes the proportion of minority to the whole society, the y-axis denotes the proportion of experiments out of n experiments that the minority has successfully affected the society, which means over 90% of agents in the society ends up with choosing action x (p_x) which is preferred by the minority agents. In Figs. 1d, e and f, the x-axis denotes the proportion of minority, and the y-axis denotes the iteration when the minority successfully influence the majority, which also represents the emergence time of the social norm in each independent experiment under the given parameters. The box-plot shows the mean value and variance of all the emergence time of the succeed experiments of the parameter set. It indicates the speed and stability of the emergence of minority influence. In Figs. 1g, h and i, the x-axis denotes the number of iterations, the y-axis denote the proportion of agents choosing action x (p_x) which is preferred by the minority agents, and each line denotes one single repeat experiment. And from left to right in each row of figures, the preference of agents are different. For other scenarios, the figures are organised in the same way.

It can be observed from Fig. 1 that with fixed preference values, as p increases, the ratio of the succeed experiments increases, and the norm emerges faster and more stable. This indicates that the larger the population of the minority, the easier the minority can affect the majority. Meanwhile, a larger k leads to a higher success ratio under the same p. Finally, the minority group with more extreme preference values can be more influential.

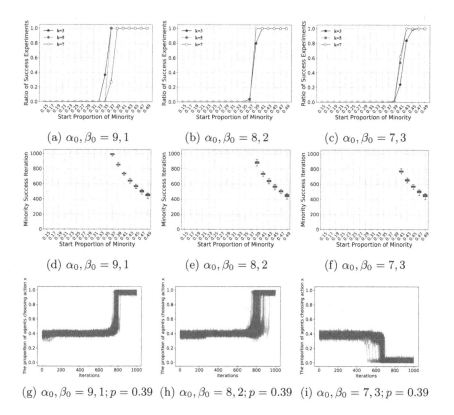

(a) $\alpha_0, \beta_0 = 9, 1$ (b) $\alpha_0, \beta_0 = 8, 2$ (c) $\alpha_0, \beta_0 = 7, 3$

(d) $\alpha_0, \beta_0 = 9, 1$ (e) $\alpha_0, \beta_0 = 8, 2$ (f) $\alpha_0, \beta_0 = 7, 3$

(g) $\alpha_0, \beta_0 = 9, 1; p = 0.39$ (h) $\alpha_0, \beta_0 = 8, 2; p = 0.39$ (i) $\alpha_0, \beta_0 = 7, 3; p = 0.39$

Fig. 2. Scenario 2-1. Neutral majority with contrary initial action

4.3   Scenario 2: Neutral Majority

Scenario 2-1: Neutral Majority with Contrary Initial Action. This scenario can be much harder, compared to the case above, for the minority to influence the majority, i.e., it requires a larger p and more iterations (Fig. 2). Consistent with Scenario 1, a higher p and a larger ratio between α_0 and β_0 can also lead to faster norm emergence. Moreover, we can observe some tipping points. For example, when we set α_0 to 9, β_0 to 1, k to 5, we can see the tipping point $p = 0.36$. This is consistent with the defection effect [13,19], which indicates that when the minority affect a few individuals of the majority, the rest of the majority can soon follow these defectors and trigger minority influence. The experimental results suggest that once p reaches the tipping point, minority influence can occur, and a social norm can emerge.

Scenario 2-2: Neutral Majority with Random Initial Action. Figure 3 shows that a relatively small minority group with extreme preference can lead the population to conform to a new social norm. This may indicate that in a society that the majority have no certain preference, a minority group with a small population can trigger the minority influence. In this scenario, we do not observe any significant correlation between k, (α_0, β_0) and the emergence of minority influence.

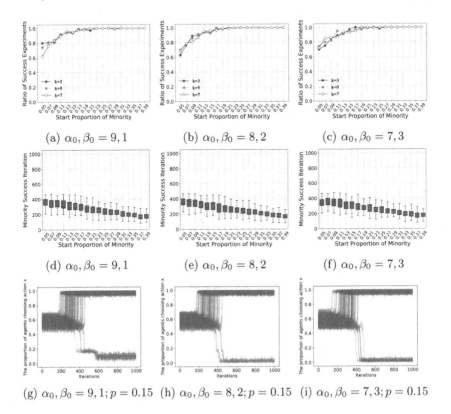

(a) $\alpha_0, \beta_0 = 9, 1$ (b) $\alpha_0, \beta_0 = 8, 2$ (c) $\alpha_0, \beta_0 = 7, 3$

(d) $\alpha_0, \beta_0 = 9, 1$ (e) $\alpha_0, \beta_0 = 8, 2$ (f) $\alpha_0, \beta_0 = 7, 3$

(g) $\alpha_0, \beta_0 = 9, 1; p = 0.15$ (h) $\alpha_0, \beta_0 = 8, 2; p = 0.15$ (i) $\alpha_0, \beta_0 = 7, 3; p = 0.15$

Fig. 3. Scenario 2-2. Neutral majority with random initial action

4.4 Scenario 3: Contrary Preference Scenario

Unlike other scenarios, we only analyse experiments with parameters $\alpha_0, \beta_0 = 9, 1, k = 7$. This is because it can be tough for the minority with gentle preference to affect the majority under this scenario. The experimental results show that minority influence occurs within 1000 iterations when p is larger than 0.45. Consistent with other scenarios, the larger the value of p, the faster the norm emerges. When p is less than 20%, we cannot observe minority influence even after 6000 iterations. This may indicate that it can be very hard for the minority to influence the majority with an opposing attitude, even when the minority is extreme and with a moderate population.

5 Conclusion and Future Works

In this paper, we proposed a novel framework, Minority Influence Model, to explore minority influence in the context of norm emergence. The model considers multiple agent interactions, where agents perform actions based on both their preferences and perceived social norm, which is reflected by gist. Based on the proposed model, extensive experiments have been done. The experimental results suggest minority influence can be correlated to the initial proportion of the minority, the extent to which the minority prefers some action, and the degree of disagreement between the minority and the majority. Furthermore, some tipping points have been observed under some scenarios. In our future work, we will extend the Minority Influence Model to structured networks [11]. We will also enrich our model to explore minority influence in more complex situations where agents can have different social positions.

References

1. Ajzen, I., Fishbein, M.: The influence of attitudes on behavior. In: The Handbook of Attitudes, vol. 173, no. 221, p. 31 (2005)
2. Brainerd, C.J., Reyna, V.F.: Gist is the grist: fuzzy-trace theory and the new intuitionism. Dev. Rev. **10**(1), 3–47 (1990)
3. Cai, Y., Zheng, H., Liu, J., Yan, B., Su, H., Liu, Y.: Balancing the pain and gain of hobnobbing: utility-based network building over attributed social networks. In: Proceedings of the 17th AAMAS, pp. 193–201 (2018)
4. Claus, C., Boutilier, C.: The dynamics of reinforcement learning in cooperative multiagent systems. In: AAAI/IAAI 1998, pp. 746–752 (1998)
5. Easley, D., Kleinberg, J., et al.: Networks, Crowds, and Markets, vol. 8. Cambridge University Press, Cambridge (2010)
6. Gardikiotis, A.: Minority influence. Soc. Pers. Psychol. Compass **5**(9), 679–693 (2011)
7. Hao, J., Leung, H.f.: The dynamics of reinforcement social learning in cooperative multiagent systems. In: Twenty-Third International Joint Conference on Artificial Intelligence (2013)

8. Hu, S., Leung, C.w., Leung, H.f., Liu, J.: To be big picture thinker or detail-oriented?: utilizing perceived gist information to achieve efficient convention emergence with bilateralism and multilateralism. In: AAMAS 2019, pp. 2021–2023 (2019)

9. Hu, S., Leung, H.: Compromise as a way to promote convention emergence and to reduce social unfairness in multi-agent systems. In: Mitrovic, T., Xue, B., Li, X. (eds.) AI 2018. LNCS (LNAI), vol. 11320, pp. 3–15. Springer, Cham (2018). https://doi.org/10.1007/978-3-030-03991-2_1

10. Kaelbling, L.P., Littman, M.L., Moore, A.W.: Reinforcement learning: a survey. J. Artif. Intell. Res. **4**, 237–285 (1996)

11. Khoussainov, B., Liu, J., Khaliq, I.: A dynamic algorithm for reachability games played on trees. In: Královič, R., Niwiński, D. (eds.) MFCS 2009. LNCS, vol. 5734, pp. 477–488. Springer, Heidelberg (2009). https://doi.org/10.1007/978-3-642-03816-7_41

12. Kittock, J.E.: Emergent conventions and the structure of multi-agent systems. In: Proceedings of the 1993 Santa Fe Institute Complex Systems Summer School, vol. 6, pp. 1–14. Citeseer (1993)

13. Levine, J.M.: Reaction to opinion deviance in small groups. Psychol. Group Influ. **2**, 187–231 (1989)

14. Liu, J., Li, L., Russell, K.: What becomes of the broken hearted?: an agent-based approach to self-evaluation, interpersonal loss, and suicide ideation. In: Proceedings of the 16th AAMAS, pp. 436–445 (2017)

15. Liu, J., Wei, Z.: Network, popularity and social cohesion: a game-theoretic approach. In: Thirty-First AAAI Conference on Artificial Intelligence (2017)

16. Milgram, S., Bickman, L., Berkowitz, L.: Note on the drawing power of crowds of different size. J. Pers. Soc. Psychol. **13**(2), 79 (1969)

17. Moscovici, S.: Innovation and minority influence. In: Perspectives on minority influence, pp. 9–52 (1985)

18. Moskvina, A., Liu, J.: Togetherness: an algorithmic approach to network integration. In: ASONAM 2016, pp. 223–230. IEEE (2016)

19. Myers, D.G.: Social Psychology, 10th edn. McGraw Hill, New York (2010)

20. Nail, P.R., MacDonald, G., Levy, D.A.: Proposal of a four-dimensional model of social response. Psychol. Bull. **126**(3), 454 (2000)

21. Nemeth, C.J.: Differential contributions of majority and minority influence. Psychol. Rev. **93**(1), 23 (1986)

22. Sen, S., Airiau, S.: Emergence of norms through social learning. In: IJCAI, vol. 1507, p. 1512 (2007)

23. Shoham, Y., Tennenholtz, M.: Emergent conventions in multi-agent systems: Initial experimental results and observations. In: Proceedings of the 3rd International Conference on Principles of Knowledge Representation and Reasoning, pp. 225–231 (1992)

24. Shoham, Y., Tennenholtz, M.: On the emergence of social conventions: modeling, analysis, and simulations. Artif. Intell. **94**(1–2), 139–166 (1997)

25. Shubik, M.: Game Theory in the Social Sciences: Concepts and Solutions, vol. 155. MIT Press Cambridge, Cambridge (1982)

26. Watkins, C.J., Dayan, P.: Q-learning. Mach. Learn. **8**(3–4), 279–292 (1992)

27. Wood, W., Lundgren, S., Ouellette, J.A., Busceme, S., Blackstone, T.: Minority influence: a meta-analytic review of social influence processes. Psychol. Bull. **115**(3), 323 (1994)
28. Yan, B., Liu, Y., Liu, J., Cai, Y., Su, H., Zheng, H.: From the periphery to the center: information brokerage in an evolving network. In: Proceedings of the 27th IJCAI, pp. 3912–3918 (2018)

A Knowledge Points and Cognitive Verb Labeling Strategy for Test Questions Based on Crowdsourcing Mode

Chonghui Guo[✉] and Meng Xu

Institute of System Engineering, Dalian University of Technology, Dalian, China
dlutguo@dlut.edu.cn, xumengwy@163.com

Abstract. Crowdsourcing, as an open knowledge production process, has the characteristics of voluntary and collaborative sharing, and has achieved remarkable results in solving many complex practical problems. In order to solve the labeling problem of knowledge points and corresponding cognitive verbs in massive test questions, this paper proposes a labeling strategy based on crowdsourcing mode. Firstly, on the basis of the three modes of crowdsourcing and the context of the problem, a crowdsourcing labeling strategy for knowledge points and cognitive verbs suitable for high school mathematics test questions is proposed. Secondly, experiments are designed and carried out to verify the feasibility of the strategy. Finally, a method of gold standard combined with adapted EM algorithm is put forward to control the quality of crowdsourcing results, and the pricing of such tasks is obtained based on experimental data.

Keywords: Crowdsourcing mode · Knowledge points labeling · Cognitive verb labeling · Gold standard data · Adapted EM algorithm

1 Introduction

Howe first proposed the concept of crowdsourcing in 2006, and crowdsourcing is a new type of problem-solving in the Internet era. It refers to the way enterprises or institutions release tasks that they cannot complete on the Internet and assign them to the general public to complete tasks [1]. Crowdsourcing results from the joint efforts of a large number of independent individuals, and at the same time it can generate more wisdom than a single individual [2]. Crowdsourcing has attracted wide attention due to its advantages of openness and low cost. Many Internet applications, such as open source software development, Wikipedia and online translation [3], have successfully applied crowdsourcing, showing significant progress beyond traditional methods.

There are a lot of data in the education industry, including test questions data, knowledge points data, students' answer records, etc. It is worth to study how to make good use of these data to better improve the quality of education. Specifically, the test questions, as the traditional carrier to examine the students' mastery of the knowledge points they have learned, are a kind of data that are worth mining. In traditional instruction scenes, students need to analyze the knowledge points involved in the test questions by themselves in the process of solving problems, and too many difficult

© Springer Nature Singapore Pte Ltd. 2019
J. Chen et al. (Eds.): KSS 2019, CCIS 1103, pp. 122–136, 2019.
https://doi.org/10.1007/978-981-15-1209-4_9

exercises are of little significance to improve the overall academic performance. So teachers need to clearly determine the design of teaching objectives in teaching process. They should be familiar with the degree to which students should master knowledge points, and not to use test questions beyond teaching objectives to examine students [4]. Furthermore, by using the test data marked with knowledge points and supplemented by the students' answer data, further researches such as learner knowledge modeling, personalized test questions recommendation, and the prediction of test difficulty can be carried out [5–7]. According to the Bloom's taxonomy, cognitive process dimension verbs marked with knowledge points can help teachers to design teaching objectives in actual teaching [8–11]. Knowledge points and cognitive verbs are the solid foundation for subsequent research, so it is particularly important to label the knowledge points and cognitive verbs involved in the test questions.

There are two classic solutions to annotation: one is to ask professional teachers to label, which is accurate but expensive. The second one is to use machine learning method completely. The learning rules are trained by the training set and the accuracy is tested by the test set. However, the labels of the training set also need to be marked manually, and the machine learning method cannot guarantee high accuracy.

In order to solve the labeling problem of knowledge points and corresponding cognitive verbs in massive test questions, this paper takes high school mathematics questions as examples, and proposes a labeling strategy based on crowdsourcing mode, crowdsourcing tasks are assigned to workers to complete jointly. Compared with professional teachers, the cost is cheaper and the precision is guaranteed and acceptable. The following sections will review related work, introduce the proposed labeling strategy, describe the experiment and then report and discuss experiment results, quality evaluation and task pricing problems in detail.

2 Related Work

2.1 Crowdsourcing

A number of studies have applied crowdsourcing in various research. In terms of crowdsourcing model, Pietro and Janis put forward three processes of evolution of human computing system: microtasking, workflow and problem-solving ecosystem, in other words, summed up three modes of crowdsourcing [12]. Pénin et al. divided online crowdsourcing into three forms: crowdsourcing for daily work, crowdsourcing for information content and crowdsourcing for innovative content [13]. Panchal believes that crowdsourcing can be divided into two types according to the competition and cooperation relationship, one is cooperation-based crowdsourcing mode, and the other is competition-based crowdsourcing mode (crowdsourcing competition) [14].

Due to different crowdsourcing tasks, problems such as task difficulty and remuneration may lead to a series of quality problems of crowdsourcing results. In the quality control of crowdsourcing products, there are a lot of related work, the most classic is the majority voting method, which assigns a task to multiple crowdsourcing workers to answer independently, then integrates the answers through voting, and regards most opinions as the final correct result. Hirth et al. used the gold standard data method in evaluating the accuracy of crowdsourcing workers [15]. However, the time and economic cost of this method are relatively high, and the evaluation method is relatively single. Iterative algorithms represented by Expectation-Maximization (EM) algorithm are also typical quality control methods. These algorithms establish models for task and crowdsourcing workers respectively, transform quality control problems into traditional problems such as maximum likelihood and optimization, and estimate worker reliability through some machine learning algorithms. On the basis of EM algorithm, many scholars have proposed an improved algorithm. Ipeirotis et al. proposed to distinguish the worker's true error rate from the worker's bias in order to improve the accuracy of the algorithm in estimating the quality of crowdsourcing workers [16]. Raykar and Yu proposed a new method to evaluate crowdsourcing workers based on EM algorithm, which can effectively distinguish malicious workers from normal crowdsourcing workers and find and replace malicious workers [17].

2.2   Bloom's Taxonomy

The applied research of Bloom's Taxonomy mainly focuses on three aspects, the first one is to evaluate the cognitive ability of research object, Aytaç and Ada studied to distinguish higher order thinking skills of the teacher candidates in the applying, analyzing, evaluating and creating categories according to the revised Bloom's Taxonomy [18]. Sharunova et al. used a cognitive game to assess the design thinking of engineering professors [19].

The second one is to support teaching activities, such as designing courses and generating test questions. Abduljabbar and Omar thought the Bloom's Taxonomy has become a common reference for the teaching and learning process used as a guide for the production of exam questions [4]. Verenna et al. used the first four levels of Bloom's taxonomy to create quiz questions designed to assess students' academic performance [20]. Arneson and Offerdahl developed an adaptation of Bloom's taxonomy specifically focused on visual representations, to aid instructors in designing instruction and assessments to target scientific visual literacy in undergraduate instruction [21]. Radmehr and Drake studied how to develop and align educational objectives, teaching activities, and assessments [22]. Amorim et al. developed a question bank on the basis of Bloom's Taxonomy and applied to a group of employees under training. The answers of employees were then used to guide the instructor's plan of teaching for the next classes [23].

The third is to use Bloom classification to classify texts and test questions. Jolliffe and Ponsford used an adaptation of Bloom's taxonomy to classify and compare mathematics questions [24]. Kusuma suggested a method that produces automation classification of Indonesian language question items based on new bloom taxonomy levels [25]. Diab and Sartawi introduced a new approach to classify the questions and learning outcome statements into Bloom's taxonomy and to verify Bloom's Taxonomy verb lists, which are being cited and used by academicians to write questions and learning outcome statements [26].

3   Proposed Strategy

In order to solve the labeling problem of knowledge points and corresponding cognitive verbs in massive test questions, this paper proposes to use crowdsourcing method to distribute crowdsourcing tasks to college students, who are the main group of crowdsourcing workers. First, formulate unified labeling rules, standardize labeling process, establish a perfect products-quality evaluation system, and screen qualified crowdsourcing workers, then distinguish high-level crowdsourcing workers from ordinary crowdsourcing ones, cooperate with high-level workers in the long term, while make general cooperation with ordinary ones, then review products of workers and pay salary to workers according to pricing strategy, at last, reward those who perform well [27]. The overall workflow is shown in Fig. 1.

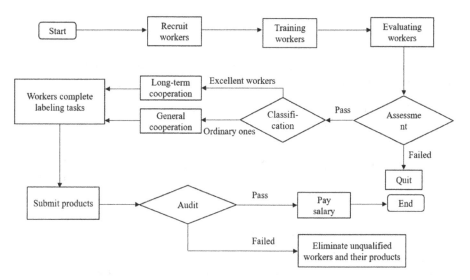

Fig. 1. The business process of task labeling for crowdsourcing workers

In the above-mentioned business process, there are four points that need special attention. This section will elaborate on how to train crowdsourcing workers to learn annotation correctly and formally, how labeling work carry out, how to evaluate and control the quality of crowdsourcing products, and the pricing problem of crowdsourcing tasks in detail below.

3.1   Training Scheme

The annotation task requires that the worker has a certain amount of subject knowledge, and at the same time, considering the cost of employment, the optimal crowdsourcing workers in this paper will be selected from college students. Crowdsourcing training has two aims. The one is to let workers understand crowdsourcing tasks and learn how to label test questions, the other one is to ensure that crowdsourcing results are available and have an elegant and standard format. Besides, there are two main subtasks in crowdsourcing tasks. The first one is at the knowledge point level: for a math problem, workers are required to label the knowledge points involved. The second one is in the cognitive dimension: for a math problem, workers are required to label cognitive verbs for corresponding knowledge points involved.

Based on above-mentioned, the training scheme proposed in this paper is as follows: firstly, recruit labeling workers from college students, then explain labeling tasks to workers, introduce background knowledge and labeling methods, demonstrate examples, and let workers practice labeling. The training process is shown in Fig. 2.

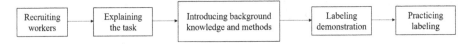

Fig. 2. Traing process

The following is an example, the question comes from the 2018 National College Entrance Test (NMET):

If the Set $A = \{x|x^2 - x - 2 > 0\}$, then $C_R A = ($).
A. $\{x|-1 < x < 2\}$ B. $\{x|-1 \leqslant x \leqslant 2\}$
C. $\{x|x < -1\} \cup \{x|x > 2\}$ D. $\{x|x \leqslant 1\} \cup \{x| \geqslant 2\}$

This question examines how to solve inequalities and the concept of complementary set, and are both at the level of comprehension, so it is labeled as follows (Table 1):

Table 1. An labeling example

Cognitive verb+Knowledge point1	Cognitive verb+Knowledge point2
Comprehend+solution of inequality	Comprehend+The concept of complementary set

3.2   Working Mode

According to the complexity of crowdsourcing tasks, the modes of crowdsourcing work could be divided into three categories. The following is a detailed discussion on which mode is suitable for this task based on the characteristics of crowdsourcing task studied in this paper.

The first mode is microtasking, as shown in Fig. 3. The complex task is divided into several small tasks, furthermore small tasks are easier to solve, thus completing the actual complex task. In the crowdsourcing task studied in this paper, the complex task is to mark out knowledge points and cognitive verbs for a large number of test questions. Each test question can be regarded as a micro task, and the complex task can be completed by finally aggregating each label.

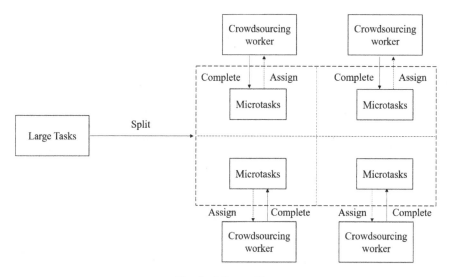

Fig. 3. Microtasking

The second mode is workflow, as shown in Fig. 4. In this mode, crowdsourcing tasks need to be evaluated first. If it is evaluated as a complex task, then it will be completed by high-level crowdsourcing workers. By contrast, if it is a simple task, it will be completed by ordinary crowdsourcing workers. Taking math questions as examples, math questions are divided into choice, gap-filling and comprehensive application questions. Generally speaking, choice and gap-filling questions are relatively simple, involving only one or two knowledge points, which can be handled by ordinary crowdsourcing workers. However, comprehensive application questions generally involve multiple knowledge points, require more knowledge, and can be given to high-level crowdsourcing workers (Fig. 5).

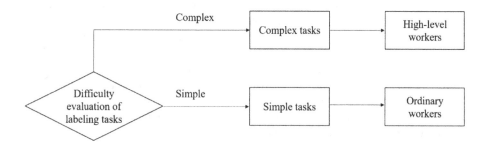

Fig. 4. Workflow

The third model is called problem-solving ecosystem, which is suitable for solving some complicated problems involving many factors. The test questions annotation is not so complicated as this, and there is no need to build such a system. However, this mode may be considered for test questions analysis and solution in the future.

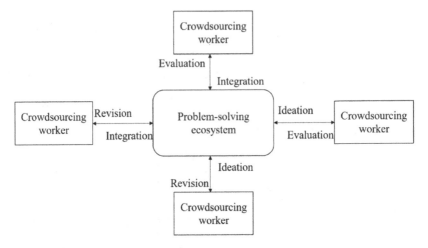

Fig. 5. Problem-solving ecosystem

In this context, the huge amount of mathematical test data to be labeled is a complicated task, indeed it is easier to divide the complicated task into micro tasks with small amount of questions and distribute them to crowdsourcing workers for labeling, in others words, microtasking complex problems. Test questions that are difficult to mark out, such as comprehensive questions, are assigned to high-level crowdsourcing workers. Test questions that are easy to label, such as choice questions and gap-filling questions, are assigned to ordinary crowdsourcing workers for completion, that is to say crowdsourcing workflow.

3.3 Quality Assessment and Control

In order to ensure the quality, first of all, we should measure the similarity of annotation products submitted by crowdsourcing workers, to figure out whether there is plagiarism. If it is qualified, then quality evaluation will be carried out.

There are three common quality evaluation schemes, the first one is gold standard data method; Second, the majority voting method; The third is EM algorithm. Gold standard data refers to a type of data with standard answers, regarded as gold standard. In this context, it means that the knowledge points of the test questions and the corresponding cognitive verbs are marked with accurate marked answers. The golden standard data method means that under the condition of standard answers, labeling workers with high accuracy are screened out, and then more tasks are given to these labeling workers, and products of these labeling workers are regarded as the final crowdsourcing results. Majority voting refers to the work results submitted by crowdsourcing workers. Each question is given a "correct" answer according to the labeling products of the majority of people, and then qualified results are screened out according to the "correct" answer to finally complete the crowdsourcing task. EM algorithm is to assume that all workers are qualified in advance, and estimate the correct annotation answer of each question by using the annotations made by multiple workers. By comparing the answers submitted by the workers with the estimated correct labeled answers, the overall error rate of each worker can be obtained, the workers with high overall error rate can be eliminated, the work products of low-quality workers can be deleted, and the high-level work results of workers with low overall error rate can be finally obtained through continuous iteration.

In this paper, it is proposed that based on common quality evaluation schemes, firstly, gold standard data are used to screen workers with guaranteed quality, so that qualified crowdsourcing workers can complete the labeling tasks without standard answers. The EM algorithm is used to iterate continuously, in each iteration, the annotation to questions were estimated according to worker reliability and products submitted by workers, and then estimate the quality of crowdsourcing workers according to the integrated answers.

The relevant definitions are as follows:

Definition 1. The test question set to be marked is defined as $I(i = 1, 2, \ldots, I)$, all knowledge point category sets are $J(j = 1, 2, \ldots, J)$, and t_{ij} is used to represent the indicator variable for labeling test question i. If the true label of i is q, $t_{ij} = 1(j = q)$, $t_{ij} = 0(j \neq q)$.

Definition 2. The probability that the real label of a test question is J is recorded as $P(j)$.

Initialization $P(j) = \frac{\sum_i t_{ij}}{Tasknum}$, $\sum_i t_{ij}$ refers to the total number of labels with a result of j, Tasknum refers to all labels provided by the workers for this test question.

Definition 3. The workers set for labeling tasks is defined as $W(w = 1, 2, \ldots, W)$, corresponding to each worker w, where n_{il}^w represents the number of times that worker W labels test question i as l, R_{jl}^w represents the probability that worker W labels tasks that were actually labeled as j as l, and R_{jl}^w is the worker's error rate matrix, the formula is as follows:

$$\left\{ R_{jl}^w \right\} = \left\{ \begin{matrix} R_{11}^w & \cdots & R_{1j}^w \\ \cdots & \cdots & \cdots \\ R_{j1}^w & \cdots & R_{jj}^w \end{matrix} \right\}$$

Definition 4. Crowdsource worker annotation set L, L is composed of l_{iw}, l_{iw} represents the annotation content made by crowdsource worker w to test question i.

The pseudocode of Adapted EM algorithm used in this paper is as follows:

Input:	Crowdsource worker's annotation set L, error threshold, iteration number c
Output:	Worker's error rate matrix R, test question labeling set T, label prior probability $P(j)$
1	Initialize an error rate matrix R of each worker.
2	Initialize the labeling set T of each test question .
3	While the error does not satisfy the given threshold and the iteration number does not reach c.
4	According to the marks made by workers on the test questions, the correct label of each test question is estimated.
5	According to the correct annotation estimated in the previous step, the error rate matrix R of crowdsourcing users is updated.
6	The prior probability $P(j)$ is estimated for the labeled label j of each test question.
7	End While
8	Return the error rate matrix R of each worker, the correct labeling set T of each test question, and the prior probability $P(j)$ of the marked label J.

The final output of the algorithm is the correct labeling set estimated from each test question and the overall error rate of each worker, and error rate is taken as a scalar value for each worker's quality evaluation, the worker with high overall error rate can be eliminated, and the products of low-level workers can be deleted. Repeated iterations eventually resulted in high-level products for workers with low overall error rate.

3.4 Pricing Strategies

After completing the labeling task within the specified time, if the final labeling products are qualified, workers can receive corresponding remuneration. The task remuneration is an important factor that affects crowdsourcing tasks. High remuneration pricing will lead to high expenditures of task initiators, while low remuneration pricing will discourage the enthusiasm of workers, thus affecting the task completion rate. From the point of view of the task initiators, the task pricing scheme should be able to ensure a higher completion rate at a lower cost. From the perspective of workers, task pricing should have certain attraction. Therefore, in the following experiment, this paper will issue questionnaires to volunteers to investigate their monthly disposable income and expected salary. Next section will discuss pricing in detail.

4   Experiment

In order to verify the feasibility of the strategy proposed in this paper, three crowdsourcing experiments were conducted among college students. Due to space constraints, this paper only introduces the latest experiment. The experiment is mainly based on math questions, with a total of 20 questions, divided into A/B test papers, each test paper has 10 math questions. These selected questions are the true questions of the college entrance test from 2012 to 2018. Most of the problems are multi-knowledge points. At the same time, a questionnaire will be issued to the volunteers so as to investigate the basic information related to this task, such as the difficulty evaluation and expected salary of the volunteers.

4.1 Labeling Products Evaluation

In the context, this paper puts forward the evaluation standard of knowledge points labeling: a test question may have multiple knowledge points, but there is only one or two main knowledge points, so as long as the main knowledge points can be labeled, the volunteer is considered to be labeled correctly. Otherwise, the volunteer fails. Besides, the evaluation standard of cognitive verb labeling is that, if the cognitive verb of the main knowledge points is labeled, then the volunteer is considered to be labeled correctly, else, the volunteer fails.

4.2 Experiment Results

A total of 35 volunteers, of whom 18 completed the task of labeling A test paper and 17 completed the task of labeling B test paper. In the end, 35 valid labeling products and 35 valid questionnaires were collected. Gold standard data refers to a type of data with standard answers [15]. This paper invites the professionals to label the knowledge points and cognitive verbs of all the experimental questions, in order to obtain the gold standard data of this task. In the context, the gold standard data are regarded as the standard annotation.

Statistics of Labeling Products
Based on the gold standard data, the average precision of labeling products of volunteers can be obtained, as shown in the following Table 2:

Table 2. The statistics of the precision of labeling products

	A	B
Knowledge point	85%	86.5%
Cognitive verb	64%	55%

It can be seen that average precision of labeling products for knowledge point is rather higher. The average precision of labeling products for cognitive verb is moderate, nevertheless, it is very close to precision performed by the machine learning method [24–26, 28], then the precision is acceptable.

Summary of Questionnaire Survey

The total number of volunteers was 35, including 32 undergraduates, accounting for 91.43%, and 3 postgraduates, accounting for 8.57%. Among the volunteers, 11 were from science and engineering, accounting for 31.43%, and 24 were from economics and management, accounting for 68.57%. Among the volunteers, 15 (42.86%) had experience as a family tutor, while 20 (57.14%) had no such experience. Among the volunteers, 7 thought the labeling task was easy, accounting for 20%, 23 thought the difficulty was moderate and acceptable, accounting for 65.71%, and 5 thought the task was very difficult and challenging, accounting for 14.29%. One of the volunteers thought that the training activities for labeling tasks had no effect and did not help labeling tasks at all, accounting for 2.86%; 14 thought that the training activities had a moderate effect and helped labeling tasks to some extent, accounting for 40%; 20 had an extraordinary effect and helped labeling tasks greatly, accounting for 57.14%. Among the volunteers, the disposable income of 8 people per month was below 1000 yuan, accounting for 22.86%, that of 9 people per month was between 1,000 and 1,500 yuan, accounting for 25.71%, that of 16 people per month was between 1,500 and 2,500 yuan, accounting for 45.71%, and that of 2 people per month was above 2,500 yuan, accounting for 5.71%. Among the volunteers, the expected salary of 8 people (Yuan/10 questions) was below 5 yuan, the expected salary of 22 people (Yuan/10 questions) was below 10 yuan, the expected salary of 28 people (Yuan/10 questions) was below 20 yuan, and the expected salary of 30 people (Yuan/10 questions) was below 30 yuan. Among the volunteers, the expected salary (yuan/hour) of 6 people was below 10 yuan, the expected salary (yuan/hour) of 14 people was below 20 yuan, the expected salary (yuan/hour) of 20 people was below 40 yuan, and the expected salary (yuan/hour) of 28 people was below 50 yuan. A total of 24 subjects took less than 20 min, 31 less than 30 min, 34 less than 40 min and 35 less than 50 min. After labeling 10 math questions, the average marking time for volunteers was 18.2 min/person.

4.3   Quality Evaluation and Control of Experimental Results

This paper proposes to screen high-level workers with gold standard and iterate the labeling products of high-level workers with adapted EM algorithm. The similarity of the labeling products provided by volunteers firstly were measured. After eliminating the suspicion of plagiarism, the labeling products of first five questions were compared with the gold standard data prepared in advance. If the correct rate reached the set threshold, the work products of these workers would be left behind. The adapted EM algorithm was used to iterate the products of the remaining five questions, and the iterated results were compared with the gold standard data.

Similarity Measure of Submitted Products

Table 3 Similarity measurement for products of A test paper

	1	2	3	4	5	6	7	8	9	10	11	12	13	14	15	16	17	18
1	1	0.1695	0.1952	0.2317	0.1827	0.2	0.2052	0.1952	0.2317	0.1866	0.1866	0.2109	0.2171	0.1907	0.2	0.1952	0.1907	0.2109
2	0.1695	1	0.2317	0.1952	0.2612	0.2109	0.1952	0.1866	0.1725	0.1791	0.2317	0.2	0.1952	0.2	0.224	0.1725	0.1907	0.2403
3	0.1952	0.2317	1	0.2109	0.2317	0.2317	0.2109	0.2	0.1907	0.1907	0.2109	0.2317	0.2109	0.2171	0.2171	0.1827	0.1952	0.2317
4	0.2317	0.1952	0.2109	1	0.25	0.2317	0.2403	0.2109	0.2612	0.2109	0.2109	0.2317	0.224	0.2171	0.2171	0.1907	0.1952	0.2317
5	0.1827	0.2612	0.2317	0.25	1	0.224	0.2317	0.2171	0.2052	0.1791	0.2317	0.224	0.2171	0.2109	0.2612	0.1866	0.2	0.224
6	0.2	0.2109	0.2317	0.2317	0.224	1	0.1952	0.1952	0.1952	0.1866	0.1952	0.2	0.1952	0.1907	0.2	0.1791	0.1757	0.2109
7	0.2052	0.1952	0.2109	0.2403	0.2317	0.1952	1	0.2109	0.2612	0.1907	0.224	0.2052	0.2403	0.2171	0.2052	0.1827	0.1866	0.2052
8	0.1952	0.1866	0.2	0.2109	0.2171	0.1952	0.2109	1	0.2109	0.1907	0.2	0.2052	0.224	0.2171	0.2171	0.2109	0.2317	0.1952
9	0.2317	0.1725	0.1907	0.2612	0.2052	0.1952	0.2612	0.2109	1	0.2109	0.2	0.2171	0.2403	0.1866	0.1866	0.1907	0.1866	0.2052
10	0.1866	0.1791	0.1907	0.2109	0.1791	0.1866	0.1907	0.1907	0.2109	1	0.1695	0.1866	0.2	0.2171	0.1725	0.1907	0.1791	0.1952
11	0.1866	0.2317	0.2109	0.2109	0.2317	0.1952	0.224	0.2	0.2	0.1695	1	0.25	0.224	0.2052	0.2317	0.1907	0.2052	0.2171
12	0.2109	0.2	0.2317	0.2317	0.224	0.2	0.2052	0.2052	0.2171	0.1866	0.25	1	0.2317	0.2	0.224	0.1866	0.2109	0.2109
13	0.2171	0.1952	0.2109	0.224	0.2171	0.1952	0.2403	0.224	0.2403	0.2	0.224	0.2317	1	0.2171	0.2171	0.2109	0.2317	0.2171
14	0.1907	0.2	0.2171	0.2171	0.2109	0.1907	0.2171	0.2171	0.1866	0.2171	0.2052	0.2	0.2171	1	0.2109	0.2171	0.2109	0.224
15	0.2	0.224	0.2171	0.2171	0.2612	0.2	0.2052	0.2171	0.1866	0.1725	0.2317	0.224	0.2171	0.2109	1	0.1952	0.2403	0.2403
16	0.1952	0.1725	0.1827	0.1907	0.1866	0.1791	0.1827	0.2109	0.1907	0.1907	0.1907	0.1866	0.2109	0.2171	0.1952	1	0.2052	0.1952
17	0.1907	0.1907	0.1952	0.1952	0.2	0.1757	0.1866	0.2317	0.1866	0.1791	0.2052	0.2109	0.2317	0.2109	0.2403	0.2052	1	0.2109
18	0.2109	0.2403	0.2317	0.2317	0.224	0.2109	0.2052	0.1952	0.2052	0.1952	0.2171	0.2109	0.2171	0.224	0.2403	0.1952	0.2109	1

Table 4 Similarity measurement for products of B test paper

	1	2	3	4	5	6	7	8	9	10	11	12	13	14	15	16	17
1	1	0.2612	0.2403	21712927	0.2109	0.2317	0.2052	0.224	0.2	0.2	0.2171	0.1952	0.25	0.2109	0.224	0.2109	0.2
2	0.2612	1	0.2612	0.2171	0.2403	0.25	0.2171	0.224	0.2403	0.2403	0.1866	0.2171	0.2743	0.2403	0.2403	0.2403	0.224
3	0.2403	0.2612	1	0.2052	0.2899	0.309	0.2052	0.2403	0.224	0.224	2052131	0.2171	0.2317	0.2403	0.2612	0.2109	0.2109
4	0.2171	0.2171	0.2052	1	0.2317	0.2	0.2109	0.2171	0.1791	0.1791	0.2	0.1907	0.2109	0.1952	0.2052	0.1791	0.2052
5	0.2109	0.2403	0.2899	23166248	1	0.2743	0.1952	0.2899	0.2109	0.2109	0.1866	0.2171	0.25	0.224	0.2109	0.2	0.2612
6	0.2317	0.25	0.309	0.2	0.2743	1	0.1907	0.2317	0.2317	0.2109	0.2109	0.2109	0.2899	0.25	0.25	0.2317	0.2052
7	0.2052	0.2171	0.2052	0.2109	0.1952	0.1907	1	0.1791	0.1791	0.1791	0.1907	0.1827	0.1907	0.1791	0.1952	0.1866	0.2052
8	0.224	0.224	0.2403	0.2171	0.2899	0.2317	0.1791	1	0.2	0.2	0.1952	0.2052	0.25	0.2612	0.2	0.1907	0.224
9	0.2	0.2403	0.224	0.1791	0.2109	0.2317	0.1791	0.2	1	1	0.1866	0.2171	0.2403	0.224	0.224	0.2612	0.1907
10	0.2	0.2403	0.224	0.1791	0.2109	0.2317	0.1791	0.2	1	1	0.1866	0.2171	0.2317	0.2403	0.224	0.2612	0.1907
11	0.2171	0.1866	0.2052	0.2	0.1866	0.2109	0.1907	0.1952	0.1866	0.1866	1	0.1757	0.2109	0.1952	0.2052	0.1952	0.1791
12	0.1952	0.2171	0.2171	0.1907	0,2171	0.2109	0.1827	0.2052	0.2171	0.2171	0.1757	1	0.2109	0.1952	0.2052	0.2052	0.2052
13	0.25	0.2743	0.2317	0.2109	0.25	0.2899	0.1907	0.25	0.2317	0.2317	0.2109	0.2109	1	0.2317	0.25	0.2317	0.2317
14	0.2109	0.2403	0.2403	0.1952	0.224	0.25	0.1791	0.2612	0.2403	0.2403	0.1952	0.1952	0.2317	1	0.224	0.2109	0.1907
15	0.224	0.2403	0.2612	0.2052	0.2109	0.25	0.1952	0.2	0.224	0.224	0.2052	0.2052	0.25	0.224	1	0.2109	0.2
16	0.2109	0.2403	0.2109	0.1791	0.2	0.2317	0.1866	0.1907	0.2612	0.2612	0.1952	0.2052	0.2317	0.2109	0.2109	1	0.1827
17	0.2	0.224	0.2109	0.2052	0.2612	0.2052	0.2052	0.224	0.1907	0.1907	0.1791	0.2052	0.2317	0.1907	0.2	0.1827	1

The abscissa and ordinate in the Tables 3 and 4 represent the corresponding volunteers. The labeling products provided by workers available from Tables 3 and 4 have very low similarity, so the suspicion of plagiarizing labeling content can be ruled out.

Knowledge Point

After adapted EM algorithm iteration, the final results of knowledge points were compared with the prepared gold standard answer. The contrast of the precision of labeling knowledge points between iterated results and volunteers' average level is as shown in Table 5:

Table 5. The contrast of the precision of labeling knowledge points

	A	B
Adapted EM	100%	80%
Average	85%	86.5%

The precision of iterated results is higher than the average level in the A test paper, while iterated results are little lower than the average level in the B test paper, but it still can maintain a high accuracy. It could be inferred that the proposed quality evaluation and control method is effective to iterate labeling knowledge points.

Cognitive Verb

If the volunteer marked out the main knowledge points of the test questions, then the volunteer's annotation would be recorded. However, if the volunteer failed to label the main knowledge points, then annotation record is considered empty. After adapted EM algorithm iteration, the final product of cognitive verbs is compared with the prepared gold standard data, and the following Table 6 can be obtained:

Table 6. The contrast of the precision of labeling cognitive verb

	A	B
Adapted EM	60%	80%
Average	64%	55%

In the A test paper, the precision of the iteration results is little lower than the average level, but the precision of the iteration results is higher than the average level in the B test paper. Therefore, it can be seen that the quality evaluation and control method is moderately effective to iterate the work products of labeling cognitive verbs.

4.4   Crowdsourcing Task Pricing

Crowdsourcing initiators and workers always hope to maximize their own interests [29]. In order to balance the interests of stakeholders, this paper mainly considers the pricing strategy from three aspects:

(1) Package and distribute more tasks to high-level workers. Select the workers with higher accuracy, exclude the workers with lower accuracy, and distribute the tasks to the workers with higher accuracy will greatly improve the task completion rate, thus avoid the problem that the tasks are delayed or even unable to be completed due to the low accuracy of the workers.

(2) Consider the monthly disposable income of workers and provide competitive salary. A total of 35 people participated in the experiment. According to the questionnaire survey, a total of 17 people have a monthly disposable income of less than 1500 yuan, which can predict that a certain number of college students or graduate students may be interested in labeling work in real life.

(3) Consider the expected salary of workers. In the previous statistics, a total of 22 people's expected salaries (RMB/10 questions) were within 10 yuan/10 questions. According to results of the questionnaire, a total of 18 people's expected salaries (RMB/hour) were within 30 yuan/hour. The previous summary also counted that a total of 24 people can complete 10 questions in 20 min, so it can be estimated that a certain number of college students or graduate students can complete 30 questions in 1 h in real labeling work and accept the price of 30 yuan/(hour * questions).

To sum up, we can set the salary of labeling tasks as 10 yuan/10 questions, and package and distribute tasks to high-level workers as far as possible.

5 Conclusion

This paper proposes a labeling strategy based on crowdsourcing mode, massive complex tasks are assigned to crowdsourcing workers to jointly complete, which can well complete the knowledge point and cognitive verb labeling tasks. The main contributions of this paper are as follows:

(1) In the field of educational data mining, crowdsourcing is proposed to solve the problem of labeling knowledge points and cognitive verbs.
(2) Based on the three modes of crowdsourcing and the context of the problem, a crowdsourcing labeling strategy suitable for subject knowledge and cognitive verbs is proposed, and an experiment are designed and carried out to verify the feasibility of the strategy.
(3) The gold standard combined with adapted EM method is proposed to control the quality of crowdsourcing products, and this method can ensure the quality of crowdsourcing products indeed.

The deficiency of this paper lies in the small scale of experiment, and the experimental results cannot accurately reflect real situations. In the future, the scale of the experiment will be expanded to continue to explore the practical application of crowdsourcing projects in the education industry.

Acknowledgements. This work was supported in part by the Natural Science Foundation of China [Grant Numbers 71771034, 71421001].

References

1. Howe, J.: The rise of crowdsourcing. Wired **14**(6), 176–183 (2006)
2. Sindlinger, T.S.: Crowdsourcing: why the power of the crowd is driving the future of business. Am. J. Health Syst. Pharmacy. **67**(18), 1565–1566 (2010)
3. Omar, F.Z., Chris, C.: Crowdsourcing translation: professional quality from non-professionals. In: Meeting of the Association for Computational Linguistics: Human Language Technologies, DBLP (2011)
4. Abduljabbar, D.A., Omar, N.: Exam questions classification based on Bloom's taxonomy cognitive level using classifiers combination. J. Theor. Appl. Inf. Technol. **78**, 447–455 (2015)
5. George, D., Jimenez, S., Baquero, J.: Automatic prediction of item difficulty for short-answer questions. In: Computing Colombian Conference. IEEE (2015)
6. Piech, C., Bassen, J., Huang, J., et al.: Deep knowledge tracing. In: Proceedings of the 28th International Conference on Neural Information Processing Systems. MIT Press (2015)
7. Milicevic, A.K., Vesin, B., Ivanovic, M., et al.: E-Learning personalization based on hybrid recommendation strategy and learning style identification. Comput. Educ. **56**(3), 885–899 (2011)

8. Bloom, B.S., Engelhart, M.D., Furst, E.J., et al.: Taxonomy of Educational Objectives, Hand-Book I: The Cognitive Domain. David McKay Co Inc., New York (1956)
9. Midgley, C.: Goals, Goal Structures and Patterns of Adaptive Learning. Routledge, New York (2014)
10. Hudak, G.M., Kihn, P.: Labeling: Pedagogy and Politics. Routledge, New York (2014)
11. Wang, Y., Sun, H.: The design of instructional objectives. Foreign Language Teaching and Research Press, Beijing (2017)
12. Pietro, M., Janis, L.D.: The power of crowds. Science **351**(6268), 32–33 (2016)
13. Pénin, J.: The limits of crowdsourcing inventive activities: what do transaction cost theory and the evolutionary theories of the firm teach us? http://cournot.u-strasbg.fr/users/osi/program/TBH_JP_crowdsouring%202010%20ENG.pdf. Accessed 11 July 2019
14. Panchal, L.Q., Panchal, J.H.: Modeling the effect of product architecture on mass-collaborative processes. J. Comput. Inf. Sci. Eng. **11**, 23–46 (2011)
15. Hirth, M., Hofeld, T., Tran-Gia, P.: Cheat-detection mechanisms for crowdsourcing. Technical report 474, University of Wurzburg (2010)
16. Ipeirotis, P. G., Provost, F., Wang, J.: Quality management on Amazon mechanical turk. In: Proceedings of the ACM SIGKDD Workshop on Human Computation, pp. 64–67. ACM (2010)
17. Raykar, V.C., Yu, S.: Ranking annotators for crowdsourced labeling tasks. In: Advances in Neural Information Processing Systems, pp. 1809–1817 (2011)
18. Aytaç, K., Ada, T.: Evaluation of mathematics teacher candidates' the ellipse knowledge according to the revised bloom's taxonomy. Univers. J. Educ. Res. **5**(10), 1782–1794 (2017)
19. Sharunova, A., Butt, M., Qureshi, A.J.: Transdisciplinary design education for engineering undergraduates: mapping of Bloom's taxonomy cognitive domain across design stages. Procedia CIRP **70**, 313–318 (2018)
20. Verenna, A.A., Noble, K.A., Pearson, H.E., et al.: Role of comprehension on performance at higher levels of Bloom's taxonomy: Findings from assessments of healthcare professional students. Anat. Sci. Educ. **11**(5), 433–444 (2018)
21. Arneson, J.B., Offerdahl, E.G.: Visual literacy in Bloom: using Bloom's taxonomy to support visual learning skills. CBE Life Sci. Educ. **17**(1), ar7 (2018)
22. Radmehr, F., Drake, M.: Revised Bloom's taxonomy and integral calculus: unpacking the knowledge dimension. Int. J. Math. Educ. **48**(8), 1206–1224 (2017)
23. Amorim, G.F., Balestrassi, P.P., Sawhney, R., et al.: Six Sigma learning evaluation model using Bloom's taxonomy. Int. J. Lean Six Sigma **9**(1), 156–174 (2018)
24. Jolliffe, F., Ponsford, R.A.: Classification and comparison of mathematics examinations—methods based on Bloom's taxonomy. Int. J. Math. Educ. Sci. Technol. **20**(5), 677–688 (1989)
25. Kusuma, S.F., Siahaan, D., Yuhana, U.L.: Automatic Indonesia's questions classification based on bloom's taxonomy using Natural Language Processing a preliminary study. In: International Conference on Information Technology Systems & Innovation. IEEE (2016)
26. Diab, S., Sartawi, B.: Classification of questions and learning outcome statements (LOS) into Bloom's taxonomy (BT) by similarity measurements towards extracting of learning outcome from learning material. Int. J. Manag. Inf. Technol. **9**(2), 01–12 (2017)
27. Guo, C.: Big Data and Protection of Ancient Villages in China. South China University of Technology Press, Guangzhou (2017)
28. Qiao, C., Hu, X.: Text classification for cognitive domains: a case using lexical, syntactic and semantic features. J. Inf. Sci. **45**(4), 516–528 (2019)
29. Mason, W.A., Watts, D.J.: Financial incentives and the "performance of crowds". ACM SIGKDD Explor. Newsl. **11**(2), 100–108 (2010)

Evaluation Algorithm for the Importance of Nodes in Directed-Weighted Networks Based on Transfer Capability Matrix

Yong Li and Xin Liu[$\boxtimes$]

Northwest Normal University, Lanzhou 730070, China
1786051081@qq.com

Abstract. Due to the heterogeneity of the complex network structure, there are some special nodes in the networks. Once these nodes fail, they will cause a large area of network crashes in a short period of time. Therefore, how to identify these special nodes in the network accurately is very important. In this paper, a node importance evaluation algorithm based on Transfer Capability Matrix is proposed. The Contribution Capability Matrix and Load Capability Matrix are defined to reflect the importance of nodes which relative to its neighbors, and then the Transfer Capability Matrix of nodes are calculated according to the above matrices. So the importance evaluation value of each node is obtained by Transfer Capability Matrix. Both considering the contribution ability and the load capacity of nodes, which makes the evaluation process more comprehensive. we also apply our method to ARPA network, Neural network, American aviation network and social network. The experiments show that our method can better mine key nodes in the networks. Finally, the effectiveness of the proposed algorithm is further verified by the cascading failures.

Keywords: Directed-weighted network · Node importance · Vital nodes · Average network efficiency

1 Introduction

In recent years, the study of complex networks has attracted widespread attention in many fields, and many excellent scientific research results have been published. The discovery of features such as the scale-free [1] and small-world characteristics [2] laid the foundation for the development of complex networks. Through the continuous exploration of scientific researchers, many important information hidden behind the networks have been discovered, which provides theoretical support for further research.

As an important research topic of complex networks, node importance assessment has wide application value. For instance, in the field of biomedicine, we can identify potential oncogenes and facilitate targeted drug use through the analysis of gene networks; by analyzing customer group relationship networks, mining "seed customers" facilitates accurate advertising and reduces the placement cost of advertisements greatly; the police identify the key members by analyzing the criminal organization's contact network, which can combat criminal gangs effectively; the network

© Springer Nature Singapore Pte Ltd. 2019
J. Chen et al. (Eds.): KSS 2019, CCIS 1103, pp. 137–148, 2019.
https://doi.org/10.1007/978-981-15-1209-4_10

administrators can extract the vital equipment in the routing network and take corresponding protective measures to improve the robustness of the network. Therefore, how to mine key nodes in the network quickly and efficiently is an important and worthy research topic [3–15].

Based on the directed weighted networks, this paper defines two node importance evaluation matrices: Contribution Capability Matrix and Load Capability matrix. According to the above index matrices, the comprehensive evaluation matrix of node importance is further calculated: Transfer Capability Matrix. Furthermore, a key node evaluation algorithm based on node transfer ability matrix is proposed. This method not only considers the load capacity of the node, but also considers the contribution ability of the node, which can better characterize the transfer ability of the nodes.

The structure of this paper is as follows: In Sect. 2, the related work is introduced; in Sect. 3, the related definitions are described, the Contribution Capability Matrix, Load Capacity Matrix and other related indicators are derived. In Sect. 4, the specific values of each index are calculated, and the Transfer Ability Matrix is obtained, then we propose the evaluation algorithm. In Sect. 5, the evaluation algorithm is applied to several specific directed-weighted networks, and the effectiveness of the proposed algorithm is verified by the cascading failure simulation experiment. In Sect. 6, this article is further summarized.

2 Related Work

In recent years, domestic and foreign scholars have proposed a number of classical node importance evaluation indicators and algorithms from different angles, including betweenness centrality [16], Katz centrality [17], PageRank [18, 19] and LeaderRank [20], etc. But these indicators and algorithms have certain limitations. The betweenness centrality is limited in practical applications. Katz centrality considers that short paths are more important than long paths. It weights paths of different lengths by a factor related to the length of the path, and calculates the number of paths by matrix inversion, which is simpler than calculating the path directly. However, due to the high time complexity, the main application is in networks with few loops and small scale. PageRank is the most classic algorithm for directed network, which is widely used in many fields. It is the core algorithm of Google search engine. PageRank defaults the probability of using the address bar to jump to other web pages is equal from any web page. However, in reality, when people browse on a boring webpage with a small amount of information, the probability of selecting a page to jump through the address bar is much greater than browsing a popular webpage with rich content. In addition, the selection of parameters in the model often requires experiments, and this parameter is not universal for different application backgrounds. Lü et al. proposed the LeaderRank algorithm, which replaces the jump probability parameter in the PageRank algorithm with a background node and the bidirectional edge of the node and all nodes in the network, which improves the accuracy of key node identification. However, the background node has the same connection with all nodes, which is inconsistent with the reality, and the method cannot be directly applied to the weighted network.

At present, there are relatively few researches on node importance evaluation methods for directed- weighted networks. Wang et al. [21] proposed the improved node importance evaluation method based on mutual information(IMI), and applied it to directed-weighted networks. This method can describe the difference between the directed-weighted network nodes in a more detailed way, but only considers the amount of information of nodes, which is obviously not comprehensive enough. Ref. [22] considered the node degree value, node efficiency and the importance contribution of adjacent nodes comprehensively, and proposed the node importance evaluation matrix(NIEM), and measured the importance degree of the node according to the dependence relationship between nodes. This method can reflect the difference of importance between nodes and can be used for key node identification of large-scale complex networks. However, it contributes the importance of the node to the adjacent nodes on average, which is inconsistent with reality.

Hu et al. [23] believed that not only the interaction between adjacent nodes, but also the non-adjacent nodes also contributed their importance to the evaluated nodes in some way, and proposed the node importance evaluation method based on the node importance contribution correlation matrix(NICCM). Ref. [24] defined the cross strength to reflect the local importance of the node; and used the total influence value of all the nodes to the evaluated node, and to characterize the relative importance of the node in the whole network. At the same time, when analyzing the influence of the influencing nodes on the evaluated nodes, not only the distance between the nodes but also the number of shortest paths are introduced; not only the effect of the influencing nodes on other nodes, but also other nodes to evaluated node are considered. Therefore, the evaluation method is more comprehensive. However, for some nodes with a degree of zero, the evaluation value is 0. Although the cross strength value is used to help distinguish the nodes, the effect is not satisfactory.

The TCM algorithm proposed in this paper calculates the Transfer Capability Matrix of the nodes based on the node Contribution Capability Matrix and the Load Capacity Matrix, and then characterizes the importance of the node. Both the weight of the connected edge between the node and its neighboring nodes is considered, and the in-strength and out-strength of the neighboring node are considered. Considering the importance of the node and the relative importance of the node to its neighboring nodes, which makes evaluation more comprehensive. And in the aspect of the complexity, universality, evaluation effect of the algorithm, it makes up for the shortcomings of the existing algorithms.

3 Related Definitions

The graph $G = \{V, E\}$ denotes a directed weighted network. $V = \{v1, v2, \cdots vn\}$ signifies the nodes set, and the edge set is $E = \{e1, e2, \cdots em\}$, $\langle i, j \rangle \in E$ indicates a directed edge from the node i to the node $j.n = |V|$ describes the number of nodes in the network, while $m = |E|$ is the number of directed edges in the network.

Definition 1. Contribution Capability Matrix(CCM)

$$CCM = \begin{bmatrix} 0 & c_{12} & \cdots & c_{1n} \\ c_{21} & 0 & \cdots & c_{2n} \\ \vdots & \vdots & \ddots & \vdots \\ c_{n1} & c_{n2} & \cdots & 0 \end{bmatrix} = \begin{bmatrix} 0 & \dfrac{\varphi_{12}}{\sum_{k=1}^{n} \varphi_{k2}} & \cdots & \dfrac{\varphi_{1n}}{\sum_{k=1}^{n} \varphi_{kn}} \\ \dfrac{\varphi_{21}}{\sum_{k=1}^{n} \varphi_{k1}} & 0 & \cdots & \dfrac{\varphi_{2n}}{\sum_{k=1}^{n} \varphi_{kn}} \\ \vdots & \vdots & \ddots & \vdots \\ \dfrac{\varphi_{n1}}{\sum_{k=1}^{n} \varphi_{k1}} & \dfrac{\varphi_{n2}}{\sum_{k=1}^{n} \varphi_{k2}} & \cdots & 0 \end{bmatrix} \quad (1)$$

Among them, φ_{ij} indicates the weight of the directed edge from node i to node j, $\sum_{k=1}^{n} \varphi_{kj}$ is the in-strength of the node j. The larger the value of c_{ij}, the greater the possibility that the node j acquire information from node i. Once the node i fails, it will have a greater impact on the acquisition of the node j, thus indicating this node is more important.

Definition 2. Load Capability Matrix(LCM)

$$LCM = \begin{bmatrix} 0 & l_{12} & \cdots & l_{1n} \\ l_{21} & 0 & \cdots & l_{2n} \\ \vdots & \vdots & \ddots & \vdots \\ l_{n1} & l_{n2} & \cdots & 0 \end{bmatrix} = \begin{bmatrix} 0 & \dfrac{\zeta_{12}}{\sum_{s=1}^{n} \xi_{1s}} & \cdots & \dfrac{\zeta_{1n}}{\sum_{s=1}^{n} \xi_{1s}} \\ \dfrac{\zeta_{21}}{\sum_{s=1}^{n} \xi_{2s}} & 0 & \cdots & \dfrac{\zeta_{2n}}{\sum_{s=1}^{n} \xi_{2s}} \\ \vdots & \vdots & \ddots & \vdots \\ \dfrac{\zeta_{n1}}{\sum_{s=1}^{n} \xi_{ns}} & \dfrac{\zeta_{n2}}{\sum_{s=1}^{n} \xi_{ns}} & \cdots & 0 \end{bmatrix} \quad (2)$$

Here, ξ_{pi} represents the weight of the directed edge from node p to node i, and $\sum_{s=1}^{n} \xi_{is}$ is the out-strength of node p. The larger the value of l_{ij}, the more likely the node i is to transfer the information for node p. Once the node i fails, it will have a greater impact on the transference of node s, thus indicating the node is more important.

Definition 3 Maximum Connectivity Coefficient [25]. According to the importance evaluation algorithm, the nodes are ranked from small to large, and observing the effect of removing a part of nodes on maximal connected subset in the network. The formula is as follows:

$$G = \frac{\gamma_{max}}{N} \quad (3)$$

Where γ_{max} is the number of nodes of the network maximal connected subset after removing a part of the nodes, N indicates the total number of nodes in the network. If the trend that the size of the network maximal connected subset becomes smaller as the nodes are removed is more obvious, which will indicate the effect of attacking the network by this method is better.

Definition 4 Network Average Efficiency [26]. It characterizes the average difficulty about circulating network information, and describes the average of the sum of the reciprocal distances of all pairs of nodes in the network.

$$E = \frac{1}{N(N-1)} \sum_{i \neq j} \frac{1}{D_{ij}} \qquad (4)$$

Here, N is the total number of nodes in the network, D_{ij} indicates the distance from node i to node j.

4 Node Importance Evaluation Algorithm Based on Transfer Capability Matrix

There are some influences between nodes in the networks, and they do not exist in isolation, especially in neighboring nodes. Therefore, this paper introduces the node Contribution Capability Matrix and Load Capability Matrix to further analyze the importance of nodes.

4.1 Evaluation Indicator Description

Figure 1 is a simple network with seven nodes. According to the formulas (1) and (2) calculating CCM and LCM respectively. The results are as follows:

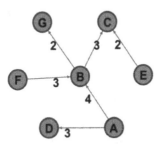

Fig. 1. A simple network.

$$
CCM = \begin{bmatrix}
0 & 0.5714 & 0 & 1 & 0 & 0 & 0 \\
0 & 0 & 0.6000 & 0 & 0 & 0 & 1 \\
0 & 0 & 0 & 0 & 0 & 0 & 0 \\
0 & 0 & 0 & 0 & 0 & 0 & 0 \\
0 & 0 & 0.4000 & 0 & 0 & 0 & 0 \\
0 & 0.4286 & 0 & 0 & 0 & 0 & 0 \\
0 & 0 & 0 & 0 & 0 & 0 & 0
\end{bmatrix}
\quad
LCM = \begin{bmatrix}
0 & 0 & 0 & 0 & 0 & 0 & 0 \\
0.5714 & 0 & 0 & 0 & 0 & 1 & 0 \\
0 & 0.6000 & 0 & 0 & 1 & 0 & 0 \\
0.4286 & 0 & 0 & 0 & 0 & 0 & 0 \\
0 & 0 & 0 & 0 & 0 & 0 & 0 \\
0 & 0 & 0 & 0 & 0 & 0 & 0 \\
0 & 0.4000 & 0 & 0 & 0 & 0 & 0
\end{bmatrix}
$$

Calculating Transfer Capability Matrix

$$TCM = CCM + LCM^T \tag{5}$$

$$TCM = \begin{bmatrix} 0 & \dfrac{\varphi_{12}}{\sum_{k=1}^{n}\varphi_{K2}} & \cdots & \dfrac{\varphi_{1n}}{\sum_{k=1}^{n}\varphi_{kn}} \\ \dfrac{\varphi_{21}}{\sum_{k=1}^{n}\varphi_{k1}} & 0 & \cdots & \dfrac{\varphi_{2n}}{\sum_{k=1}^{n}\varphi_{kn}} \\ \vdots & \vdots & \ddots & \vdots \\ \dfrac{\varphi_{n1}}{\sum_{k=1}^{n}\varphi_{k1}} & \dfrac{\varphi_{n2}}{\sum_{k=1}^{n}\varphi_{k2}} & \cdots & 0 \end{bmatrix} + \begin{bmatrix} 0 & \dfrac{\zeta_{12}}{\sum_{s=1}^{n}\xi_{1s}} & \cdots & \dfrac{\zeta_{1n}}{\sum_{s=1}^{n}\xi_{1s}} \\ \dfrac{\zeta_{21}}{\sum_{s=1}^{n}\xi_{2s}} & 0 & \cdots & \dfrac{\zeta_{2n}}{\sum_{s=1}^{n}\xi_{2s}} \\ \vdots & \vdots & \ddots & \vdots \\ \dfrac{\zeta_{n1}}{\sum_{s=1}^{n}\xi_{ns}} & \dfrac{\zeta_{n2}}{\sum_{s=1}^{n}\xi_{ns}} & \cdots & 0 \end{bmatrix} \tag{6}$$

$$TCM = \begin{bmatrix} 0 & \dfrac{\varphi_{12}}{\sum_{k=1}^{n}\varphi_{k2}} + \dfrac{\zeta_{21}}{\sum_{s=1}^{n}\xi_{2s}} & \cdots & \dfrac{\varphi_{1n}}{\sum_{k=1}^{n}\varphi_{kn}} + \dfrac{\zeta_{n1}}{\sum_{s=1}^{n}\xi_{ns}} \\ \dfrac{\varphi_{21}}{\sum_{k=1}^{n}\varphi_{k1}} + \dfrac{\zeta_{12}}{\sum_{s=1}^{n}\xi_{1s}} & 0 & \cdots & \dfrac{\varphi_{2n}}{\sum_{k=1}^{n}\varphi_{kn}} + \dfrac{\zeta_{n2}}{\sum_{s=1}^{n}\xi_{ns}} \\ \vdots & \vdots & \ddots & \vdots \\ \dfrac{\varphi_{n1}}{\sum_{k=1}^{n}\varphi_{k1}} + \dfrac{\zeta_{1n}}{\sum_{s=1}^{n}\zeta_{1s}} & \dfrac{\varphi_{n2}}{\sum_{k=1}^{n}\varphi_{k2}} + \dfrac{\zeta_{2n}}{\sum_{s=1}^{n}\xi_{2s}} & \cdots & 0 \end{bmatrix} \tag{7}$$

Calculating the comprehensive evaluation indicator

$$C_i = \sum_{j=1}^{n} \dfrac{\varphi_{ij}}{\sum_{k=1}^{n}\varphi_{kj}} + \dfrac{\zeta_{ji}}{\sum_{s=1}^{n}\xi_{js}} \tag{8}$$

5 Empirical Analysis

5.1 Algorithm Effectiveness Analysis

We choose the ARPA network, Neural network, USAir97 network and Karate network to verify the effectiveness of our algorithm. Although this network is an undirected-unweighted network, we turns it into a directed-weighted network by weighting method [27] and directing method [21], as shown in Fig. 2.

Table 1 lists the node importance ranking results of the ARPA network with three methods.

Firstly, it can be seen from Table 1. The top five important nodes are 2,3,19,6,14 with our method and the other nodes except node 6 belong to the union of the top five node sets acquired by NICCM and NIEM. which reflects the effectiveness of TCM. In this paper, the node 6 ranks 4th, while Ref. [22] ranks in the thirteenth, and in Ref. [23] ranks in the sixth. the worse the connectivity of the network after deleting the nodes,

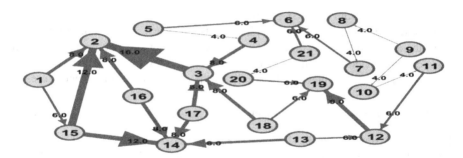

Fig. 2. The directed-weighted ARPA network.

Table 1. The node importance ranking results of the network in Fig. 2.

TCM		NICCM	NIEM
Values	Node ID	Node ID	Node ID
2.5714	2	3	2
2.4351	3	14	3
2.4286	19	2	15
2.2000	6	12	14
2.0714	14	19	17
2.0000	9	6	16
1.8571	12	4	18
1.5000	8	13	19
1.5000	10	15	13
1.5000	11	18	12
1.3333	5	5	4
1.3333	7	11	1
1.3333	21	20	6
1.1818	1	17	20
1.0346	15	7	5
0.7333	4	10	11
0.7222	13	21	21
0.6857	20	8	7
0.6190	18	16	10
0.5556	17	9	8
0.4040	16	1	9

the more effective the algorithm is. Therefore, based on the importance evaluation results under different evaluation algorithms in Table 1, we compare the scale of the largest subgraph and the number of subgraphs by removing the top 5 important nodes of ranking results. And find that the more subgraphs will indicate worse connectivity when nodes are removed. The smaller the scale of the largest subgraph is, the worse the network connectivity becomes. So the corresponding key node identification method is more accurate. The experimental results are shown in Fig. 3.

It can be seen from the first picture in Fig. 3 that when the top five nodes are removed from the node importance ranking results of Table 1, our method can result in more subgraphs and the scale of largest subgraph is less than others obviously. We can also find similar phenomena in the other three pictures. Therefore, our method is better than others, and it can achieve better results in the evaluation of node importance.

5.2 Further Verify the Algorithm Through Cascading Failure Simulation

The above experimental results show the effectiveness of TCM algorithm. However, in order to further verify its reliability, this paper analyzes the robustness of ARPA network, Neural network, Karate club network and 1997 North American aviation network USAir97. Where the number of subgraphs after failure and the maximum connectivity coefficient are used as the metrics. The cascading failure process is as follows: According to various evaluation methods, we can obtain the importance ranking results of all nodes in the network, and then the nodes and corresponding edges are deleted in turn according to the sorting results. The number of subgraphs S is larger, the smaller the maximal connectivity coefficient G is, which will indicate the robustness of the network is worse and the corresponding evaluation algorithm is more accurate. Figures 4, 5, 6 and 7 are the results of the cascading failure simulation experiments corresponding to our method, NICCM and NIEM.

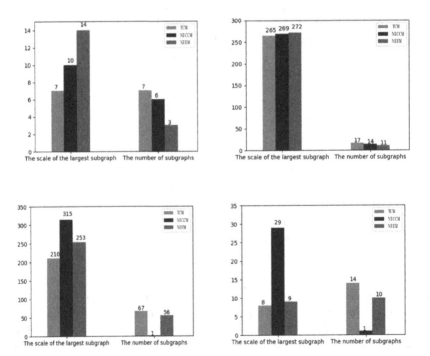

Fig. 3. The scale of the largest subgraph and the number of subgraphs by removing the top k important nodes.

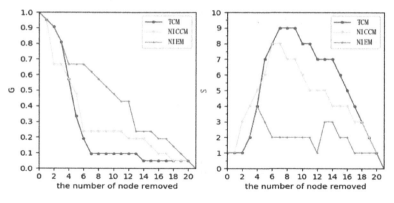

Fig. 4. The robustness comparison under different evaluation methods on ARPA networks.

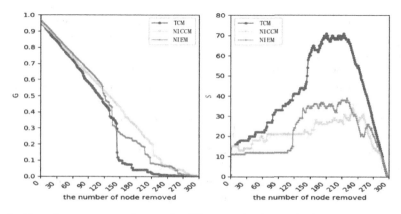

Fig. 5. The robustness comparison under different evaluation methods on Neural networks.

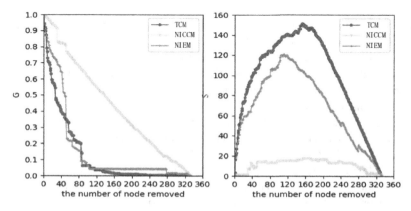

Fig. 6. The robustness comparison under different evaluation methods on USAir97 networks.

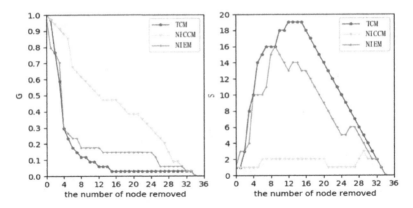

Fig. 7. The robustness comparison under different evaluation methods on Karate club networks.

We found that when the number of removed nodes increases, our method can cause a larger decline. For the ARPA network, although the performance of deleting the second, and third nodes is temporarily behind NICCM, the algorithm is significantly better than the other two algorithms after deleting the top 5 nodes. The maximal connectivity coefficient of our method is smaller than that of NICCM and NIEM. In addition, it can be seen from the trend of change that the number of subgraphs corresponding to our algorithm is the same as NIEM, but slightly lower than NICCM after removing the fourth node. When the 5th node is removed, however, the value is always ahead of other methods. Observing the experimental simulation of the USAir97, Neural and Karate Club networks, the results of the proposed algorithm are significantly better than NICCM and NIEM. The above experiments show that the algorithm TCM proposed in this paper is relatively more reliable in identifying key nodes in the network.

6 Conclusion

This paper uses the nodes CCM and LCM reflects the importance of the node relative to its adjacent nodes indirectly. Then we propose a node importance evaluation algorithm based on Transfer Capability Matrix(TCM), which can work well for directed-weighted networks. The Space Complexity and the Time Complexity of our algorithm is $O(n^2)$. The feasibility of the algorithm is verified by a number of typical directed-weighted networks.

The study found that the TCM algorithm can distinguish the importance of the ARPA network more closely. When removing some important nodes obtained by the algorithm, it can make the network average efficiency drop rapidly. And when the top five key nodes are removed, the number of subgraphs can be more and the scale of the largest subgraph is smaller, which indicates TCM algorithm can affect the connectivity of the network significantly. So the algorithm has good reliability.

In this paper, the cascading failure effects of ARPA network, Neural network, Karate club network and USAir97 network are simulated. It is found that TCM algorithm can reduce the maximum connectivity coefficient G and make the number of subgraphs S larger overall, which verifies the reliability of the algorithm. Therefore, the proposed algorithm can be widely applied to the node importance evaluation of directed-weighted networks. It provides a theoretical basis for further exploration of complex networks. How to calculate transfer capability matrix more accurately will be the next research work.

References

1. Barabási, A.L., Albert, R.: Emergence of scaling in random networks. Science **286**(5439), 509–512 (1999)
2. Watts, D.J., Strogatz, S.H.: Collective dynamics of 'small-world' networks. Nature **393**, 440–442 (1998)
3. Poulin, R., Boily, M.C., Mâsse, B.: Dynamical systems to define centrality in social networks. Soc. Netw. **22**(3), 187–220 (2000)
4. Ren, Z.M., Shao, F., Liu, J.G., Guo, Q., Wang, B.H.: Node importance measurement based on the degree and clustering coefficient information. Acta Phys. Sin. **62**(12), 522–526 (2013)
5. Liu, Y.H., Jin, J.Z., Zhang, Y., Xu, C.: A new clustering algorithm based on data field in complex networks. J. Supercomput. **67**(3), 723–737 (2014)
6. Yu, H., Liu, Z., Li, Y.J.: Key nodes in complex networks identified by multi-attribute decision-making method. Acta Phys. Sin. **62**(02), 54–62 (2013)
7. Fan, W.L., Hu, P., Liu, Z.G.: Cascading failure model in power grids using the complex network theory. IET Gener. Transm. Distrib. **10**(15), 3940–3949 (2016)
8. Han, Z.M., Wu, Y., Tan, X.S., Duan, D.G., Yang, W.J.: Ranking key nodes in complex networks by considering structural holes. Acta Phys. Sin. **64**(05), 429–437 (2015)
9. Han, Z.M., Chen, Y., Li, M.Q., Liu, W., Yang, W.J.: An efficient node influence metric based on triangle in complex networks. Acta Phys. Sin. **65**(16), 289–300 (2016)
10. Liu, J.G., Lin, J.H., Guo, Q.: Locating influential nodes via dynamics-sensitive centrality. Sci. Rep. **6**, 21380 (2016)
11. Morone, F., Makse, H.A.: Influence maximization in complex networks through optimal percolation. Nature **524**(7563), 65–68 (2015)
12. Lin, J.H., Guo, Q., Dong, W.Z.: Identifying the node spreading influence with largest k-core values. Phys. Lett. A **378**(45), 3279–3284 (2014)
13. Liu, Y., Tang, M., Zhou, T.: Improving the accuracy of the k-shell method by removing redundant links: From a perspective of spreading dynamics. Sci. Rep. **5**, 13172 (2015)
14. Wen, X.X., Tu, C.L., Wu, M.G., Jiang, X.R.: Fast ranking nodes importance in complex networks based on LS-SVM method. Phys. A: Stat. Mech. Appl. **506**, 11–23 (2018)
15. Xu, S., Wang, P.: Identifying important nodes by adaptive LeaderRank. Phys. A.: Stat. Mech. Appl. **469**, 654–664 (2017)
16. Freeman, L.C.: A set of measures of centrality based on betweenness. Sociometry **40**(1), 35–41 (1977)
17. Katz, L.: A new status index derived from sociometric analysis. Psychometrika **18**(1), 39–43 (1953)
18. Brin, S., Page, L.: The anatomy of a large-scale hypertextual Web search engine. Comput. Networks. Isdn. **30**(1–7), 107–117 (1998)
19. Radicchi, F., Fortunato, S., Markines, B., Vespignani, A.: Diffusion of scientific credits and the ranking of scientists. Phys. Rev. E **80**(5), 056103 (2009)
20. Lü, L.Y., Zhang, Y.C., Yeung, C.H., Zhou, T.: Leaders in social networks, the delicious case. PLoS ONE **6**(6), e21202 (1999)

21. Wang, B., Ma, R.N., Wang, G., Chen, B.: Improved evaluation method for node importance based on mutual information in weighted networks. J. Comput. Appl. **35**(7), 1820–1823 (2015)
22. Zhou, X., Zhang, F.M., Li, K.W., Hui, X.B., Wu, H.S.: Finding vital node by node importance evaluation matrix in complex networks. Acta Phys. Sin. **61**(05), 1–7 (2012)
23. Hu, P., Fan, W.L., Mei, S.W.: Emergence of scaling in random networks. Physica A: Stat. Mech. Appl. **429**, 169–176 (2015)
24. Wang, Y., Guo, J.L.: Evaluation method of node importance in directed-weighted complex network based on multiple influence matrix. Acta Phys. Sin. **66**(05), 19–30 (2017)
25. Ruan, Y.R., Lao, S.Y., Wang, J.D., Bai, L., Chen, L.D.: Node importance ranking of complex network based on degree and network density. Acta Phys. Sin. **66**(03), 371–379 (2017)
26. Latora, V., Marchiori, M.: A measure of centrality based on network efficiency. New J. Phys. **9**, 188 (2007)
27. Mirzasoleiman, B., Babaei, M., Jalili, M., Safari, M.: Cascaded failures in weighted networks. Phys. Rev. E Stat. Nonlinear Soft Matter Phys. **84**(4 Pt 2), 046114 (2011)

Opinion Dynamics Considering Social Comparison in Online Social Networks

Mengmeng Liu[(✉)] and Lili Rong

Institute of Systems Engineering,
Dalian University of Technology, Dalian 116023, China
liummengdj@163.com

Abstract. Social comparison theory holds that people will evaluate their opinions by comparison with someone close to their own abilities or opinions. Inspired by this, we propose an opinion dynamic model referring to classic Deffuant model combined with online communication characteristics. Opinions in this model are produced along with information diffusion and thus result in dynamic population in opinion interactions. And interactions are allowed between commenters who comment on the same post or repost. Through simulations, our model is observed to present more divergent opinions than classic Deffuant model as is also believed in reference [15]. Moreover, effects of different compromise thresholds, namely confidence bound, and comparison thresholds are discussed. Further simulations are also conducted to clarify the impacts of diffusion parameters on opinion evolution.

Keywords: Opinion dynamics · Social comparison · Compromise theory · Online social networks

1 Introduction

Opinion dynamics have attracted extensive attention in multiple disciplines such as psychology [1], physics [2, 3], computer science [4] since the proposal of Ising model [5]. Especially with the emergence of various social platforms, opinion interactions in online social networks arouse keenly interests of scholars worldwide.

Previous researches have shed light on the mechanism of how individuals interact with each other on activities like election or discussions on social events [6, 7]. Most of existing models describing these phenomena will end up into convergence [8] and thus form consensus. For discrete models with finite choices, such as voter model [4, 9], majority rule model [10], Sznajd model [11], convergence is more easier to achieve due to limited states of each individual. Differently, as continuous models in which opinions are distributed randomly between 0 and 1, Deffuant model [12, 13] and HK model [14] absorb the compromise theory when define rules for interactions. In Deffuant model, individuals interact with their neighbors when differences of their opinions are smaller than a threshold which is also called the confidence bound, while in HK model individuals update their opinions taking references from all their neighbors with opinions falling under the threshold.

© Springer Nature Singapore Pte Ltd. 2019
J. Chen et al. (Eds.): KSS 2019, CCIS 1103, pp. 149–159, 2019.
https://doi.org/10.1007/978-981-15-1209-4_11

Interactions are merely allowed between neighbors as a common practice in these dynamic models, which may work in more private social networks, such as Tencent QQ, Wechat, Facebook. In more public or open social networks like microblogs, alternative means for interactions between users are provided. Particularly in sina weibo, one of the largest social platforms in China, users' comments are visible to others and commenting on the same original post makes it possible for commenters to communicate with each other. Usually information posted by media accounts with a large number of followers will attract quite a few comments. Interactions among such amount of users may not always result in opinion convergence to average opinion as in previous models, which is also mentioned in [15, 16]. As Buunk [17] puts it, social comparison is a central feature of human social life. Social comparison theory is first proposed by Festinger [18] focusing on ability and opinions. Based on the social comparison theory, people evaluate their opinions by comparison with opinions of others; and someone close to one's own ability or opinion will be chosen for comparison [18]. As a result, people tend not to conform to the average observed response but to exceed it in the socially preferred direction [19].

As mentioned in References [20, 21], Deffuant model is better suited for pairwise interactions within large population. Despite of this, individuals produce their opinions online gradually after receiving certain information which is different from that in Deffuant model, in which opinions are assigned to each individual initially. Therefore, on the basis of Deffuant model, characteristics of communication in online social networks need to be introduced.

Based on previous researches and characteristics of opinion interactions in online social networks, following issues need to be figured out: (1) Differences of evolution results under scenarios with dynamic and static populations in interactions (2) Impacts of compromise and comparison boundaries on final opinion distribution (3) Effects of certain information diffusion on opinion evolution results. In order to reveal these, we propose an opinion dynamic model considering compromise behavior combined with social comparison phenomena in opinion interactions. In this model, individuals not only interact with their neighbors but also other commenters and with the diffusion of relevant information, individuals constantly join in the interactions. The remainder of the paper is structured as follows. Section 2 presents the opinion dynamic model. In Sect. 3 we conduct simulation experiments and analysis. Finally Sect. 4 presents the conclusions.

2 Proposed Model

2.1 Stages of Opinion Dynamic Process

From a macro perspective, propagation process of information or opinion online is commonly divided into several periods, which are based on the density calculation of different states of individuals. Similarly, single individuals will come through several stages in opinion dynamics as well. According to opinion dynamics studied previously, we divide the opinion evolution process from perspective of single individual into three stages: initial opinion formation stage, modification stage and stability stage.

In formation stage, individuals receive information of a topic or issue and produce their own opinions. After formation, they may change their opinions as a result of interaction with others on the basis of personal perception. Then, if no other information keeps involving in, opinions will reach a stable state. So in our work, evolution process will be investigated considering only single information (Fig. 1).

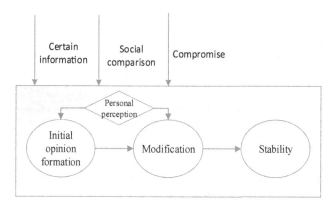

Fig. 1. Opinion dynamic process

2.2 Communication Network Through Comments in Online Social Networks

As is widely acknowledged, information or opinion is exchanged among neighbors. While for many online social platforms featured with openness, the communications could be far more than that. In Fig. 2, each node represents users in microblogs and links reflect relationships among them; nodes in yellow are users spreading the information. Dotted lines stand for weak ties among users; nodes in green and orange represent nodes who comment on different posts respectively.

When hub nodes release information, their followers will receive the message in certain probability and then comment on it. All users could view any comments benefiting from the openness, so weak ties are built among them, which provides chances for them to communicate with each other even between strangers (users with no links between them). Interaction network for node P in Fig. 2(b) is taken as an example for newly formed interaction pattern. From the aspect of network topology, a global coupling network composed by commenters is formed due to information spreading of node Q (as an example) in Fig. 2(c).

2.3 Model Description

Generally, users online may get information from others who have reposted or posted it. If the information is worth discussing, public opinions will arise. Opinion evolution can be related with information diffusion in the sense that when users repost the

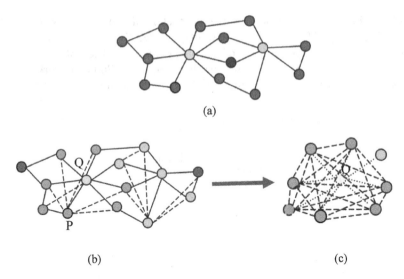

(a)

(b)   (c)

Fig. 2. Communication network. (a) Sketch map of information diffusion network; (b) Opinion interaction network; (c) Global coupling network

information with no opinions, their neighbors may produce their opinions (comments) after receiving it. Especially in social events, various opinions will be presented due to its public-concern characteristic. Based on the above analysis, an opinion dynamic model referring to online social networks is proposed as in Fig. 3.

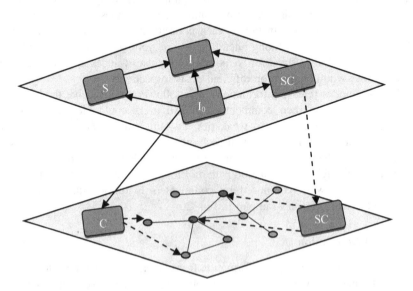

Fig. 3. Two-layered opinion dynamics model with information diffusion

The first layer in Fig. 3 illustrates the information diffusion process, and the second layer depicts opinions communications. Directed lines in solid black represent the transformations among different states. At each time step, individuals are in one of these possible states:

Ignorant (I_0): individuals who are not exposed to the information;
Informed (I): informed state, represents individuals who received the information with no potential behaviors;
Spreader (S): individuals who repost the information;
Commenter (C): individuals who produce their opinions by comments;
Spreader&Commenter (SC): individuals who propose opinions by reposts attached with opinions.

As shown in Fig. 3, Ignorants may be exposed to certain information when their neighbors are in active state of spreading the information. Some of them may transfer into Informed ones with the probability of λ based on the mean-field theory and they will neither repost the information nor produce any opinions later.

In online social networks like sina weibo, twitter, users can take part in the discussion either by comments, reposts or reposts with comments attached. Thus, individuals in state I_0 will transfer into Commenters, Spreaders, or Spreader & Commenter with probability α, β, η, respectively. Here, we assume that there is no conversion from C to SC, which indicates that Spreaders will not continue to comment on the information once they have made their choices at first. As new reposts continue, there is a chance σ that previous ones may be covered, thus lose the ability to spread and become informed again.

With the spreading of information, Ignorants will constantly become aware of the information and join the interactions, which brings about the dynamic population in opinion interactions. The directed links from state C indicate two users who comment on the same post or repost of their neighbor in color red; Another two directed links from SC represent two users with both commenting and reposting behaviors. As mentioned above, these two types of connections will result in discrepant ways of interactions. Inspired by classic Deffuant model and social comparison theory, opinion interactions rules can be expressed as follows.

a. For nodes in state SC: At each step, a link is randomly chosen; If both ends are in SC state, they may interact with each other. According to Deffuant model, the interactions are limited by the confidence bound. People tend to compare with similar others and become more extreme than others referring to social comparison theory.
Let node i and node j be the interaction pairs, $O_i, O_j \in (0, 1]$.

$$\begin{cases} O_i(t+1) = O_i(t) + \mu\left[O_j(t) - O_i(t)\right] \\ O_j(t+1) = O_j(t) + \mu\left[O_i(t) - O_j(t)\right] \end{cases}, \varepsilon_1 < |O_j(t) - O_i(t)| < \varepsilon \qquad (1)$$

Where, ε_1 denotes the upper limit for individuals to compare with others holding close opinions, above which they will change their opinions in the light of compromise theory. ε is the confidence bound.

$$\begin{cases} O_i(t+1) = O_i(t) + \mu\left[O_j(t) - O_i(t)\right] \\ O_j(t+1) = O_j(t) + \mu\left[O_i(t) - O_j(t)\right] \end{cases}, \left|O_j(t) - O_i(t)\right| < \varepsilon_1, O_{i/j}(t) \in (0.4, 0.6) \quad (2)$$

Inspired by 5-point Likert scale, opinions can be divided into five levels between (0, 1), in which opinions in (0.4, 0.6) are regarded as moderate ones with no obvious preference. Therefore, Eq. (2) defines interaction rule under situations when either one or both present moderate opinions.

$$\begin{cases} O_i(t+1) = random(O_i(t) + \mu\left[O_j(t) - O_i(t)\right]) \\ O_j(t+1) = random(O_j(t) + \mu\left[O_i(t) - O_j(t)\right]) \end{cases}, \left|O_j(t) - O_i(t)\right| < \varepsilon_1, O_{i/j} \in (0, 0.4]$$

$$(3)$$

$$\begin{cases} O_i(t+1) = O_i(t) + \mu\left[O_j(t) - O_i(t)\right] + random(1 - O_i(t) - \mu\left[O_j(t) - O_i(t)\right]) \\ O_j(t+1) = O_j(t) + \mu\left[O_i(t) - O_j(t)\right] + random(1 - O_j(t) - \mu\left[O_i(t) - O_j(t)\right]) \end{cases} \quad (4)$$

Individuals are posited to become more extreme than before when compare with similar opinions based on social comparison. Equation (4) holds under the condition where the interaction pairs have similar tendencies under constraints of $\left|O_j(t) - O_i(t)\right| < \varepsilon_1$ when $O_{i/j} \in [0.6, 1]$.

b. For nodes following nodes in state SC: if there is only one end in SC state of the link, the other one may change its opinion following rules above with node in SC remains unchanged.

c. For nodes in state C: An indicator *source* is introduced to identify whether these individuals comment on the same post. Interactions may occur if commenters comment on the same post since all comments under a post will be listed below (Fig. 2), otherwise there is little chance for them to see each others' comments besides neighbors.. At each step, two nodes in state C with are randomly chosen; If they do, their opinions will exchange based on Eqs. (1)–(4) as well.

3 Simulation Experiments and Analysis

3.1 Parameters Settings

As in classic Deffuant model, individuals will produce their opinions randomly in (0, 1). The probability of either reposting or commenting is posited to be larger than both. Parameters are set as $\mu = 0.5$, $\varepsilon_1 = 0.2$, $\varepsilon = 0.5$, $\lambda = 0.1$, $\alpha = 0.2$, $\beta = 0.2$, $\eta = 0.1$, $\sigma = 0.01$.

The value of *Source* of nodes is set to be the id of node whom they receive the information from. Since there is access limitation for checking others' followers in sina weibo, a directed sub-network of twitter composed on 6 ego-networks is extracted from Stanford Large Network Dataset [22] in this paper due to the similarity of the two social platforms. Degree distributions of the sub-network are shown in Fig. 4 and the basic topological properties is listed in Table 1.

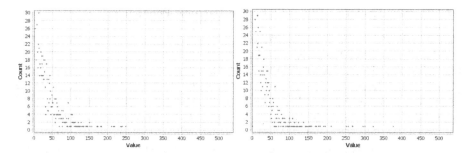

Fig. 4. (a) In-degree distribution; (b) Out-degree distribution

Table 1. Metrics of the network

Nodes	Edges	Clustering coefficient	Average path	Average degree
1108	33370	0.394	3.11	30.117

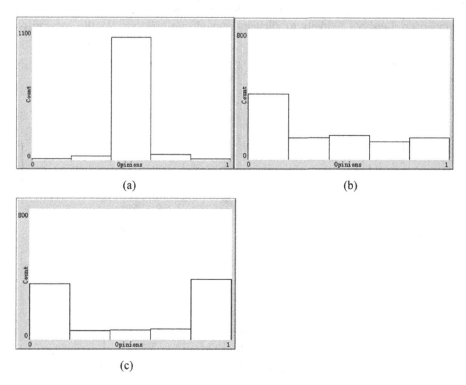

(a)

(b)

(c)

Fig. 5. Final opinion distributions under different scenarios; (a) Classic Deffuant model; (b) Opinion interactions only between neighbors in proposed model; (c) The proposed model

3.2   Opinion Evolution Under Static and Dynamic Population

Opinions in this model are produced with the information diffusion other than pre-distributed, therefore the opinion attached to a repost can not affect others when the repost loses its ability to spread. In this section, evolution processes are simulated to compare the difference between final opinion distributions of Deffuant model and the proposed model. As shown in Fig. 5, compared to classic model, the proposed model exhibited a more divergent result when only considering interactions between neighbors due to the dynamic population and the limited ability of spreading in online social network. Interactions between strangers (i.e. commenters who comment on the same repost or post) through comments will promote the communications and lead to a more extreme result.

3.3   Effects of Varying Social Comparison and Compromise Thresholds on Opinion Evolution

In order to find out how thresholds of comparison and compromise affect opinion evolution. Simulations are carried out under varying ε_1 and ε, and the final opinions are distributed as in Fig. 6. In this model, larger compromise threshold with smaller comparison threshold is likely to elicit more moderate opinions. As the comparison

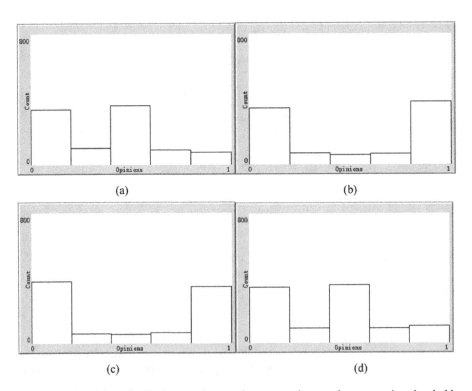

Fig. 6. Final opinion distributions under varying comparison and compromise thresholds. (a) $\varepsilon = 0.5$, $\varepsilon_1 = 0.1$; (b) $\varepsilon = 0.5$, $\varepsilon_1 = 0.25$; (c) $\varepsilon = 0.4$, $\varepsilon_1 = 0.2$; (d) $\varepsilon = 0.4$, $\varepsilon_1 = 0.08$

threshold increases, public opinion unfolds an extreme result. Opinions are found to be more sensitive to comparison threshold, which reveals why extreme opinions show up frequently in discussions over some issues. Therefore, guiding public opinions to be more moderate is possible by focusing on how to enlarge the compromise threshold, for example releasing immediate evidence-based information.

3.4   Effects of Diffusion Parameters on Opinion Evolution

Section 3.2 elaborates the discrepant results under different ways of interactions, in which interactions between neighbors and communications among commenters may result in different opinion distributions. Simulations are designed to investigate how diffusion parameters affect the opinion dynamics. Results in Fig. 7 point out that, the final distributions of opinions will not change greatly when more users choose not to react to the information. While changing the probability of commenting or reposting behaviors will exert an obvious impact on the results. Since interactions among commenters are not strictly constrained by the topology, more commenters are likely to result in more extreme opinions. When users tend to repost the information with comments attached, the final opinion distribution is observed to be more divergent due to the limited spreading time.

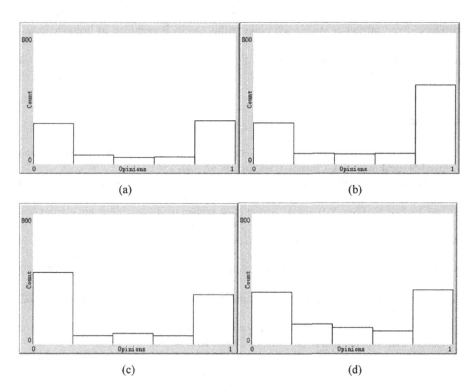

Fig. 7. Final opinion distributions. (a) $\lambda = 0.2$, $\alpha = 0.2$, $\beta = 0.2$, $\eta = 0.1$; (b) $\lambda = 0.1$, $\alpha = 0.3$, $\beta = 0.2$, $\eta = 0.1$; (c) $\lambda = 0.1$, $\alpha = 0.2$, $\beta = 0.3$, $\eta = 0.1$; (d) $\lambda = 0.1$, $\alpha = 0.2$, $\beta = 0.2$, $\eta = 0.2$

4 Conclusions

Online social networks represented by mircroblogs play a significant role in information spreading and opinion dynamics. Opinion interactions in these social networks like sina weibo are not limited to neighbors, thus provides an alternative way of communication. In this paper, opinion dynamics in online social networks are investigated introducing social comparison on the basis of Deffuant model with considering information diffusion and interactions between commenters.

Relevant information diffusion leads to dynamic population in discussion which results in different opinion distributions with classic Deffuant model finally. When transformation probability from Ignorants to Informed state increases with other parameters remaining unchanged, no obvious difference is observed besides changes in densities of individuals participating in the interactions. Besides, effects of thresholds for compromise based on Deffuant model and comparison on opinion evolution are discussed; Simulation results indicate that smaller comparison threshold combined with larger confidence bound may bring out moderate opinions, while with the small increment of comparison threshold, opinions will obviously turn extreme. Diffusion parameters are found to function significantly in final opinion distributions as well by changing the scale of users in different states.

Though our model is developed in terms of communications in sina weibo, it reveals the common characteristics in many other platforms. Further studies will introduce users' attributes since each node in this model is regarded as homogeneous.

Acknowledgments. This work is supported by the National Natural Science Foundation of China under Grant Nos. 71871039, 71871042, 71421001.

References

1. Jang, K., Park, N., Song, H.: Social comparison on Facebook: its antecedents and psychological outcomes. Comput. Hum. Behav. **62**, 147–154 (2016)
2. Li, M., Dankowicz, H.: Impact of temporal network structures on the speed of consensus formation in opinion dynamics. Physica A **523**, 1355–1370 (2019)
3. Lee, E., Lee, S., Eom, Y.H., et al.: Impact of perception models on friendship paradox and opinion formation. Phys. Rev. E **99**(5), 052302 (2019)
4. Urena, R., Kou, G., Dong, Y., et al.: A review on trust propagation and opinion dynamics in social networks and group decision making frameworks. Inf. Sci. **478**, 461–475 (2019)
5. Glauber, R.J.: Time dependent statistics of the Ising model. J. Math. Phys. **4**(2), 294–307 (1963)
6. Lorenz, J.: Continuous opinion dynamics under bounded confidence: a survey. Int. J. Mod. Phys. C **18**(12), 1819–1838 (2007)
7. Zhao, Y., Zhang, L., Tang, M., et al.: Bounded confidence opinion dynamics with opinion leaders and environmental noises. Comput. Oper. Res. **74**, 205–213 (2016)
8. Kuhlman, C.J., Kumar, V.S.A., Ravi, S.S.: Controlling opinion propagation in online networks. Comput. Netw. **57**(10), 2121–2132 (2013)
9. Sood, V., Redner, S.: Voter model on heterogeneous graphs. Phys. Rev. Lett. **94**(17), 178701 (2019)

10. Krapivsky, P.L., Redner, S.: Dynamics of majority rule in two-state interacting spin systems. Phys. Rev. Lett. **90**(23), 238701 (2003)
11. Sznajd-Weron, K., Sznajd, J.: Opinion evolution in closed community. Int. J. Mod. Phys. C **11**(06), 1157–1165 (2000)
12. Deffuant, G., Neau, D., Amblard, F., et al.: Mixing beliefs among interacting agents. Adv. Complex Syst. **3**(01n04), 87–98 (2000)
13. Shang, Y.: Deffuant model with general opinion distributions: first impression and critical confidence bound. Complexity **19**(2), 38–49 (2013)
14. Hegselmann, R., Krause, U.: Opinion dynamics and bounded confidence models, analysis, and simulation. J. Artif. Soc. Soc. Simul. **5**(3), 1–33 (2002)
15. Tian, Y., Wang, L.: Opinion dynamics in social networks with stubborn agents: an issue-based perspective. Automatica **96**, 213–223 (2018)
16. Parsegov, S.E., Proskurnikov, A.V., Tempo, R., et al.: Novel multidimensional models of opinion dynamics in social networks. IEEE Trans. Autom. Control **62**(5), 2270–2285 (2017)
17. Buunk, A.P., Gibbons, F.X.: Social comparison: the end of a theory and the emergence of a field. Organ. Behav. Hum. Decis. Process. **102**(1), 3–21 (2007)
18. Festinger, L.: A theory of social comparison processes. Hum. Relat. **7**(2), 117–140 (1954)
19. Myers, D.G.: Polarizing effects of social comparison. J. Exp. Soc. Psychol. **14**(6), 554–563 (1978)
20. Castellano, C., Fortunato, S., Loreto, V.: Statistical physics of social dynamics. Rev. Mod. Phys. **81**(2), 591 (2009)
21. Dong, Y., Zhan, M., Kou, G., et al.: A survey on the fusion process in opinion dynamics. Inf. Fusion **43**, 57–65 (2018)
22. McAuley, J., Leskovec, J.: Learning to discover social circles in ego networks. In: Proceedings of the 26th Annual Conference on Information Processing Systems, Lake Tahoe, USA, pp. 539–547 (2012)

Research on Dissemination Value of Micro-Blog Information and Empirical Study

Liangliang Li[1,2], Yijun Liu[1,2(✉)], Yuxue Chi[1,2], and Ning Ma[1]

[1] Institutes of Science and Development, CAS, Beijing 100190, China
yijunliu@casipm.ac.cn
[2] University of Chinese Academy of Sciences, Beijing 100049, China

Abstract. Microblog has become one of the most important platforms for public opinion propagation of social hot events. Studying the information dissemination value on microblogs has profound significance for the response and governance of public opinions. Based on the Microblog platform, this paper uses social network analysis method, innovatively proposes emotional PageRank value to measure the affective commitment of a microblog, and from three dimensions to build an indicator system to evaluate the dissemination value of a microblog. Finally, a comparative analysis is made on the dissemination value of important microblogs concerning a hot topic, and the factors affecting the dissemination value of microblogs will be discussed.

Keywords: Node influence · Capacity of edge transmission · Network cohesion · Emotional PageRank algorithm · Dissemination value

1 Introduction

The value of information can only be realized through effective information dissemination. In epidemic information, for instance, the information dissemination may alert individuals. In the real-world conditions, once someone gets infected, his friends who get the information may take actions to reduce their risks to be infected by him and his intimates [1]. In disaster information, effective information dissemination plays a critical role in decreasing response time and reducing the number of deaths and economic losses [2]. In marketing communications, electronic or online word of mouth (eWOM) has become an important factor in consumer buying decisions [3]. Opinion leaders' eWOM influences product sales through product experience effects and knowledge background effects [4]. Many companies receiving favorable eWOM have a better chance to increase sales [5]. The proliferation of social network sites (SNSs) has promoted new forms of social interaction of information [6]. Online social media, such as WeChat and microblog in China, has become the significant information platforms for people to obtain and share information. Interconnections between people on social network sites enhance the process of information dissemination and amplify the influence of that information [7]. Thus, how to measuring the influence of the information, as well as extracting the value of information dissemination from social networking sites, becomes a meaningful work.

According to Statistical Reports on Internet Development in China, as of December 2018, netizen usage rate of microblog is up to 42.3%, with the annual growth of 10.9%.

© Springer Nature Singapore Pte Ltd. 2019
J. Chen et al. (Eds.): KSS 2019, CCIS 1103, pp. 160–174, 2019.
https://doi.org/10.1007/978-981-15-1209-4_12

Through the microblog, people interact and express personal experiences, perspectives and opinions more easily. Microblog can quickly generate timely and diverse public opinion information sources after the occurrence of social hot events. By reposting, commenting and other propagation modes, the platform aggregates broad public opinion information such as netizen's perception, attitudes, opinions, which contain huge dissemination value. Internet public opinion is the main topic of social governance, while the openness, timeliness and vitality of microblog make it of great significance to effectively extract dissemination value from public opinion information on microblog to guide and intervene in network public opinion, so as to serve the society better.

The remaindering sections of this paper are organized as follows: Sect. 2 reviews related works of indicators of information dissemination value in social network sites. Section 3 emphasizes the emotional PageRank algorithm. Section 4 will introduce an indicator system including the method, concept and model to evaluate the dissemination value of microblog information, while Sect. 5 applies it to an actual data set. The paper concludes with a discussion providing some suggestions for further research work.

2   Related Work

Research on models of measuring information dissemination value in social network sites were scarce. Particularly, impact maximization in the field of social network analysis is related to finding the most influential nodes. Consequently, individuals at central or critical positions in the microblogging networks are regarded as influential and are expected to play a vital role in spreading information. Relatedly, Sun and Tang [8] focused on measuring the influence levels of nodes and edges qualitatively and quantitatively in the network. Gao found that betweenness centrality best explains the influence on information dissemination [9]. Up to now, a lot of related work has been done on measuring the influence of a node. For example, node influence can be measured based on node degree (Degree centrality [10]), shortest path (Closeness centrality [11] and betweenness centrality [12]) and random walk (eigenvector centrality [13], Katz centrality [14] and PageRank algorithm [15]). Medley found that most of the central nodes can sometimes link to a particular person [16], which proved that the bridge node may be an important node to maximize the network cohesion force, so the in-degree and out-degree indicators might not sufficient to recognize the most influential node in the information dissemination process [17]. In addition, PageRank algorithm is the most common metric for measuring nodes' influence. At present, most pertinent literature calculating nodes' PageRank value and sorting social network opinion leaders' influence are based on the network topology [18], which is lack of emotional indicators(the emotional PageRank algorithm is innovatively introduced in Sect. 3 in response to this problem).

In view of the situation that in social network, messages from a node can spread to other nodes, yet measuring the node influence of a user inevitably holds some limitation in measuring the dissemination value of a blog. Relevant scholars have emphasized the potential of edges or structure for controlling of communication in social networks. For example, 19. Girvan et al. believed that the key edges play roles in supporting nodes [19]. Freeman proposed edge betweenness, to effectively identify the influential edges [20]. In terms of social network structure, Watts believed that the best network with better information communication should be a network with a short average distance between nodes and a high degree of clustering [21].

In summary, the analysis of information dissemination value based on online public opinion is vital on current social network research. All the above indexes can quantitatively measure the influence of nodes, the importance of connected edges or the cohesiveness in the information communication on the social network, but there is a lack of relevant literature on the dissemination value of microblogs. In addition, there is a lack of quantitative calculation on the affective commitment in microblogs to better measure the node influence of blogs. The paper will make contributions to the literature by introducing an innovative, targeted and accurate information transmission value index system based on the users' reposting behavior on microblog considering that the main approaches in information spread in microblog is reposting.

3   Preliminarily: The Emotional PageRank Algorithm

As explained above, influencial nodes in a network play a vital role in spreading information. When calculating node influence, it often considers structure of the network using social network analysis method. However, a node with large amounts of in-degree doesn't means influential if it receives negative comments mostly. So, it's necessary to innovatively introduce the emotional PageRank algorithm.

3.1   Research on PageRank Algorithm

This section will discuss the research of PageRank algorithm. Based on the random walk model, PageRank algorithm calculates the importance of pages on the Web using centrality measures [22]. As a variety of eigenvector centrality, PageRank manifests that when a node with high centrality is linked to many other nodes, all these nodes will also acquire high centrality. Since the PageRank algorithm evaluating the importance of web pages was proposed, relevant scholars have modified the PageRank algorithm to identify opinion leaders in social networks. For example, based on supernetwork theory, Ma et al. proposed a SuperedgeRank algorithm to detect opinion leaders [23]. Lü et al. proposed a ground node based on PageRank and devised LeaderRank (LR), which is parameter free and eliminates the calibration steps of PageRank [24]. Based on blogosphere, Song et al. considered network structure and its information novelty of blogosphere to propose InfluenceRank for true opinion leader identification.

The downside of these modified algorithm is that they only consider the topology and location in the network, which does not reflect the propagation value of a microblog, especially when one receive amounts of negative comments. Subsequently, considering negative opinions, TwitterRank et al. [25–27] were proposed for identifying opinion leaders. Regarding to negative links, the Sim-PR [28], Vir-PR [29] and PT [30] were proposed to identify potential opinion leaders from online comments. Chen considered both positive and negative opinions to propose TrustRank [31], yet identified the post sentiment orientation only, which was not precise.

3.2   Emotional PageRank Algorithm

Based on above discussion, this paper innovatively uses emotional lexicon to calculate the emotional score of a microblog.

Sentiment lexicons play an important role in sentiment analysis tasks, which include opinion words, sentiment phrases, and idioms with sentiment polarities [32]. Huang et al. proposed a method to identify emotions of internet public sentiment based on Chinese emotional lexicon and obtained good results in empirical research [33]. There are a total of 1200 high frequency emotional words occuring in microblogs as well as 50 emotional buzzwords in the emotional lexicon, scored on a scale of 1 (negative) to 9 (positive) according to their average sentiment.

Because of the domain-specific in sentiment expression, this paper updates Huang's emotional lexicon. Firstly, based on a software platform named *Public Opinion Collection and Mining* in Chinese Academy of Sciences, we collect some emotional words using the Chinese word segmentation technology. Secondly, we expands Huang's emotional lexicon with some new buzzwords such as "hhhhhh" and "laugh till cry". Then we use Delphi method to determine their emotional scores. Finally, the emotional lexicon in this paper has 2565 words with 767 positive words, 1671 negative words and 127 neutral words.

The proposed emotional PageRank algorithm combined emotional analysis and weighting method. To measure the emotional weight of a repost from node j to node i, the repost will be broken up into words with the methods of word segmentation, and the words will be matched with the updated emotional lexicon. The corresponding emotional scores will be added to the scores of the repost if the words are matched.

Similarly, the emotional weight W_{ij} can be written as Eq. 1.

$$W_{ij} = \frac{\sum e_{ij}}{n_{ij}} \tag{1}$$

Where W_{ij} is the average emotional scores of a repost from node j to node i, $\sum e_{ij}$ refers to all of the emotional scores of reposts from node j to node i, and n_{ij} is the number of forwarding blogs from node j to node i.

The emotional PageRank algorithm can be written as Eq. 2.

$$PR(i) = \frac{1 - \alpha}{N} + \alpha \sum_{j \in R(i)} PR(j) \frac{W_{ij}}{\sum_{k \in T(j)} W_{kj}} \tag{2}$$

Where $PR(i)$ is the emotional PageRank scores of node i, and α is the damping factor which is 0.85 in general. N refers to the total number of the nodes in forwarding network. $R(i)$ is the set of nodes point to i, and $\sum_{k \in T(j)} |W_{kj}|$ is the absolute value of the weights of the edges linked from j.

4 Method, Concept and Model

4.1 Social Network Analysis

The advantages of the social network analysis (hereinafter: SNA) in understanding various social phenomenon are obvious [34]. SNA has a long history of application in human sociology [35, 36], where the earliest can be traced back to 1994 [37]. As an interdisciplinary methodology, SNA uses nodes and edges to represent social networks, map the relationships and flows between people, groups, organizations and other connected information entities, which can quantitatively analyze the complex information transmission and interaction in social networks on the basis of mathematics and graph theory. In the specific social context, SNA focuses on the actors and their relationships to understand networks and their participants [38]. In business applications, SNA can identify user preferences. In governance of global terrorism, SNA has shown the potential to effectively analyze cyber terrorist communities [39, 40]. With its effectiveness, high operability and great application value, SNA becomes one of the simplest and most mature methods to study the information transmission of social network.

Hence, this paper uses SNA to study the dissemination value of microblog posts based on users' reposting behavior because the main approaches in information spread in microblog is reposting. Let's define $G = <V, E>$ as a directed network where $V = \{v1, v2, v3, ...\}$ is a set of nodes and $E = \{e1, e2, e3...\}$ is the set of edges. Letting M represents all microblog sets in the network and R the forwarding relationship of M, we consider the triplet $WB = (G, M, R)$ as the object of research. This paper define dissemination value as DV. Analogously, the contributions of nodes, connecting edges and network structure to the dissemination value of a blog are defined respectively as node influence (NI), capacity of edge transmission (CET) and network cohesiveness (NC). The evaluation model of dissemination value of microblogs information is shown as Eq. 3.

$$DV = f(NI, ETC, NC) \tag{3}$$

4.2 Node Influence

The influence of a node in social networks refers to the ability to influence the information transmission of other nodes [41], Nodes with high influence have important contributions to the dissemination value of information.

Node influence is mainly based on the closeness of the relationship between nodes. Meanwhile, influential nodes in social network are important nodes with high social centrality and high reputation [42]. Therefore, based on reposting behavior in microblogs, this paper uses degree of centricity and betweenness centrality to measure node influence due to their representativeness. As a high positive forwarding is an important factor of node influence, this paper creatively use the emotional PageRank algorithm introduced above to measure the identity and credibility of nodes. The measurable indicators of node influence are shown in Table 1.

Table 1. Measurable indicators of node influence

Indicator	Notion	Formula	Practical significance
Degree of centricity (C_D)	The degree to which a node is connected to other nodes in a network	$C_D(v) = \frac{d_v}{|N|-1}$	The importance and influence of a node in a network
Betweenness centrality (C_B)	The proportion of the shortest path from node s to node t which pass by node v	$C_B(v) = \sum\limits_{s \neq v \neq t \in V} \frac{\sigma_{st}(v)}{\sigma_{st}}$	The importance of a node as a bridge
Emotional PageRank algorithm (EPR)	The emotional identification values of all nodes using the average emotional score of the reposting text as the initial emotional identification degree	$EPR(i) = \frac{1-\alpha}{N} + \alpha \sum\limits_{j \in R(i)} PR(j) \frac{W_{ij}}{\sum_{k \in T(j)} W_{kj}}$	Emotional identification of a node

4.3 Capacity of Edge Transmission

In the social network, weak ties have already become the important information bridge for nodes to get new resources, which are found mainly between nodes that contact infrequently. These edges with large capacity of transmission make chain propagation possible among different clusters in the network. Therefore weak ties with large capacity of transmission in the network make the information dissemination more efficiently and cost effectively [43].

However, the present documents don't embody capacity of edge transmission in social networks. Therefore, this paper proposes capacity of edge transmission (CET) to measure the power of an edge to spread information in the social network. Breadthways, the capacity of edge transmission can be measured by edge betweenness. While lengthways, it can be measured by the average forwarding level of the edge, that is, the higher the average forwarding level is, the larger the promulgation scope is, and so as the dissemination value. The average forwarding level of a blog can be written as Eq. 4.

$$L_F = \frac{f_1 + 2f_2 + \ldots + nf_n}{|E|} \tag{4}$$

Where n refers to the max forwarding level of the blog and $f_1, f_2, \ldots, f_n$ refer to the forwarding level of 1, 2,... n respectively. The average forwarding level of a blog should be divide by the overall edges in the reposting network of the blog.

The measurement indicators of capacity of edge transmission are shown in Table 2.

Table 2. Measurable indicators of capacity of edge transmission

Indicator	Notion	Formula	Practical significance
Average edge betweenness (B_E)	The proportion of the shortest paths from node s to node t which pass by edge e_{ij}	$B_E(e_{ij}) = \dfrac{\sum_{e_{ij} \in E} \sum_{s,t \in V} \frac{\sigma(s,t|e)}{\sigma(s,t)}}{|E|}$	The ability to transmit and control network resources
Average forwarding level (L_F)	The average forwarding level of the reposting network of a blog	$L_F = \dfrac{f_1 + 2f_2 + \ldots + nf_n}{|E|}$	Transmit deepness

4.4 Network Cohesion

Only studying the nodes and edges in the network cannot reveal the dissemination value of a blog from the perspectives of network structure. Lewin (1930s) first pointed out that cohesion was the centripetal force for members of a group. Network cohesion is the degree of mutual attraction and close relation between nodes in the network. The higher the network cohesion is, the greater is the degree of intimacy and affinity for nodes, which can promote the harmonious development of network propagation.

In a broad sense, cohesion refers to the stickiness of a group, which can be understood as the power source to promote the sustainable development of the group for group members [44]. Many researchers of group dynamics agreed that cohesion is a key feature of successful groups [45, 46]. In the field of social network analysis, cohesion is also related to intimate relationships, because relationships with similar actors are often transitive [47].

Network cohesion can be measured from two perspectives. One is to measure the overall network cohesion by network density. The other is to capture the local connections and interrelationships within substructures and the connectivity between cohesive groups within the network [48]. Therefore, network density and clustering coefficient are used to measure network cohesion in this paper.

However, relevant studies also found that smaller groups can nurture more trust, commitment and cohesion [49]. The underlying reason may be that smaller groups have more opportunities to interact and communicate with each other [50]. In addition, the analysis of team performance shows that teams with smaller size have stronger cohesion and therefore perform better [51]. Presumably, they also apply to real-life social groups. Therefore, in order to better measure the cohesion of the network, this paper uses the network diameter to measure the size of the network. The larger the network diameter is, the larger the network is, which is negatively correlated with the network cohesion. The measurable indicators of network cohesiveness are shown in Table 3.

Table 3. Measurable indicators of network cohesiveness

Indicator	Notion	Formula	Practical significance
Network density (ρ)	A ratio between the number of actual edges and the maximum number of possible edges in a network	$\rho(g) = \dfrac{M}{N(N-1)}$	The sparsity or denseness of a network structure
Network diameter (D)	The maximum distance between any two nodes in a network	$D(g) = \min(d_{st})$ $(s \neq t, s, t \in V)$	The size of the network
Clustering coefficient (CC)	A node's clustering coefficient is the ratio between actual edges of the node's neighbor node and the maximum number of them. The clustering coefficient of a network is the average of all node's clustering coefficients	$CC_i = \dfrac{e_i}{k(k-1)}$ $CC = \dfrac{\sum_1^N CC_i}{N}$	The degree of information asymmetry

4.5 Indicator System

To sum up, we have constructed indicator system to evaluate the dissemination value of microblogs, which is shown in Fig. 1.

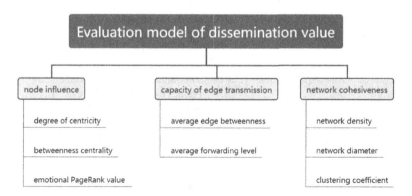

Fig. 1. The evaluation model of dissemination value

Where the node influence, the capacity of edge transmission and the network cohesiveness are formulated respectively as Eqs. 5, 6 and 7.

$$NI = f(C_D(v), PR(v), C_B(v)) \tag{5}$$

$$ETC = f(L_F(e), B_E(e)) \tag{6}$$

$$NC = f(\rho(g), D(g), CC(g)) \tag{7}$$

5 Computational Experiments

5.1 Data Set

With great relevance to Sino-US Trade War, 5G has become a hot topic recently. Moreover, most countries along the Belt and Road have a great demand for information services so that 5G technology has great advantages in improving their communication environment. In a word, 5G has aroused a wave of public opinion.

We collected all the microblogs concerning 5G and filter out blogs with less reposts to figure out the most important events about 5G. We identified six key microblogs on the basis that: (a) they were in pace with the crucial events concerning 5G, (b) they received a great many reposts among several similar blogs, and(c) the number of reposts of these blogs are similar, which guarantees the comparability of the result and high reliability of the results analysis. In the whole public opinion information on 5G topic, these blogs are representative with high volume of reposts and also suitable for the empirical study. The blogs are shown as Table 4, with amount of more than 3000 total reposts involves.

Table 4. Data set

Node	User categories	Date	Main content of Micro-blog posts	Forwarding number
A	Information center of the ministry of industry and information technology	June 26, 2019	China has officially entered the first year of 5G commercial development	604
B	Professional newspaper	June 7, 2019	The initial price of 5G mobile phones exceeds 10,000 yuan	564
C	Electric businessman	June 28, 2019	5G mine remote control engineering vehicle	636
D	Huawei terminal official microblog	July 22, 2019	China's first huawei 5G dual-mode mobile phone to receive a license for 5G terminal telecom equipment	629
E	Sina finance official microblog	July 24, 2019	5G base station radiation is negligible	664
F	Migu video technology co. LTD	June 27, 2019	The first 8K cinema in the country	529

We will analyze the dissemination value of these microblogs from some aspects and make suggestions on how to improve the dissemination value of microblogs based on the evaluation model proposed above.

The network structure of the 6 blogs are shown as Fig. 2, where (a), (b), (c), (d), (e) and (f) represent A, B, C, D, E and F respectively.

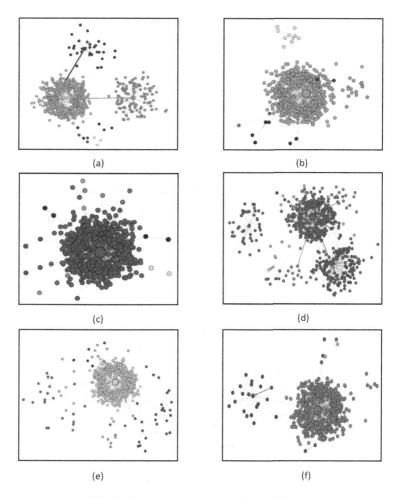

Fig. 2. The network structure of the 6 blogs.

5.2   Results

The node influence, capacity of edge transmission and network cohesiveness of the blogs are shown as Tables 5, 6 and 7, respectively.

After normalization, the results are shown as Table 8. The overall *NI*, *CET*, *NC* and *DV* results are shown as Table 9.

Table 5. Node influence (*NI*)

Node	C_D	C_B	EPR
A	0.753	0.031	6.036E-09
B	0.948	0.004	4.136E-09
C	0.984	0.319	4.156E-09
D	0.594	0.043	1.058E-08
E	0.911	0.006	6.701E-09
F	0.946	0.007	7.916E-09

Table 6. Capacity of edge transmission (*CET*)

Node	B_E	L_F
A	0.003	1.252
B	0.002	1.066
C	0.008	1.016
D	0.003	1.437
E	0.002	1.140
F	0.002	1.057

Table 7. Network cohesiveness (*NC*)

Node	ρ	D	CC
A	1.721E-03	3	3.377E-03
B	1.834E-03	4	3.604E-03
C	1.615E-03	2	1.592E-02
D	1.756E-03	4	1.936E-02
E	1.604E-03	4	3.384E-03
F	2.112E-03	2	6.140E-03

Table 8. Results after normalization

	C_D	C_B	EPR	B_E	L_F	ρ	D	CC
A	0.409	0.086	0.295	0.087	0.560	0.230	0.5	0
B	0.907	0	0	0.002	0.118	0.454	1	0.014
C	1	1	0.003	1	0	0.023	0	0.785
D	0	0.124	1	0.192	1	0.300	1	1
E	0.814	0.006	0.398	0		0.295	1	0.00043
F	0.903	0.010	0.587	0.034	0.097	1	0	0.173

Table 9. The *NI*, *CET*, *NC* and *DV*

	NI	CET	NC	DV
A	0.790	0.647	−0.270	1.167
B	0.907	0.121	−0.532	0.496
C	2.003	1.000	0.808	3.811
D	1.124	1.192	0.300	2.616
E	1.218	0.295	−1.000	0.514
F	1.499	0.131	1.173	2.802

5.3   Results Analysis

The results shows that C's degree of centricity is highest in Table 8, because there are no important nodes except C, and the forwarding is almost to C, which can be easily seen from Fig. 3. The betweenness centrality of C is the highest, too. They all imply C the influence and the importance degree as a bridge in a network. However, the emotional PageRank algorithm results show that D is the highest. As shown in Table 4, D is about upcoming China's first huawei 5G dual-mode mobile phone receiving a license for 5G terminal telecom equipment, which arouses strong patriotism and great expectations for the future among netizens. B has the lowest scores in emotional PageRank algorithm results, because B's post is about the initial price of 5G mobile phones, from which most netizens are not going to get their hopes up. The average edge betweenness of E is the smallest, mostly because E has the largest amounts of edges. D has the highest forwarding level due to some influencial nodes in the forwarding network, which can obviously be seen from Fig. 2.

As can be seen in Table 9, on node influence, C ranks No. 1, followed by the F, E, D, B, A. It's because for node C, the degree of centricity and betweenness centrality is the highest. In terms of capacity of edge transmission, D take the top spot followed by C, A, E, F and B, because in D's forwarding network, there are at least two influencial nodes and the edges in the network are averagely important in transmit and control network resources, also, the transmit deepness is high overall. On network cohesiveness, F is the highest owing to the smallest size of the network and better degree to which adjacent nodes form a group, which makes the information from node F more symmetrical.

The order of dissemination value of information in the six blog is C, F, D, A, E, B. We can see that except C, the others are all organizations or companies. So, it can be inferred that the most dissemination value may come from some ordinary "grass", which provides some new ideas on public sentiment guidance. Moreover, the two smallest dissemination value comes from E and B, and E is about the initial price of 5G mobile phones while B informs that 5G base station radiation is negligible, the former post is disappointing while the later arouses suspicion. It can be seen that our evaluation model of dissemination value can reflect the communicating effect of online public opinion to a certain degree.

To sum up, the reposts from influencial microblog users can improve the dissemination value of microblogs. And the transmission value of a blog is reflected not only in the amount of reposts, but also in the emotional attitude of the reposts. Moreover, weak ties reduce the cost of the information dissemination in microblogs and faster the propagation speed, which can be obtain from microblog users not familiar with each other. Multistage forwarding of a blog deepens transmission depth and enables a more profound dissemination value. In microblog network, users with frequent interaction manifest network cohesiveness and can absorb, transform and spread information easilier and better.

6 Discussion

It is of great practical significance to study the dissemination value of online public opinions on microblogs for the intervention and guidance of public opinions. The research on the information dissemination value of online social network is rare, and there is no quantitative model to measure the dissemination value of microblog posts. Based on SNA, this paper creatively proposes an evaluation model of dissemination value on microblog from the perspectives of node influence, capacity of edge transmission and network cohesiveness. Most innovative, this paper proposes the calculation method of emotional PageRank value, which can effectively measure the emotional identification of a blog, therefore implies the dissemination value of the blog.

However, the research in this paper is based on analyzing the forwarding behavior of information dissemination value on microblogs, while the dissemination value also come from other behaviors, such as commenting, giving likes and so on. Therefore, this paper has reference significance for the future research based on other communication behaviors. From the other perspective, it's of great value to increase amount of dataset. Finally, this paper studies the dissemination value of microblogs based on SNA, and cannot describe the frequent interaction between users over time, which is the important aspect to invest in the future.

References

1. Lu, Y., Liu, J.: The impact of information dissemination strategies to epidemic spreading on complex networks. Phys. A: Stat. Mech. Appl. 120920 (2019)
2. Zhang, N., Huang, H., Su, B.: Comprehensive analysis of information dissemination in disasters. Phys. A: Stat. Mech. Appl. **462**, 846–857 (2016)
3. Guernsey, L.: Suddenly, Everybody's an Expert on Everything. The New York Times, New York (2000)
4. Bao, T., Chang, T.L.S.: Finding disseminators via electronic word of mouth message for effective marketing communications. Decis. Support Syst. **67**, 21–29 (2014)
5. Chevalier, J.A., Mayzlin, D.: The effect of word of mouth on sales: Online book reviews. J. Market. Res. **43**(3), 345–354 (2006)
6. Bakshy, E., Hofman, J.M., Mason, W.A., Watts, D.J.: Everyone's an influencer: quantifying influence on Twitter. In: Proceedings of the fourth ACM International Conference on Web Search and Data Mining, pp. 65–74. ACM (2011)
7. Luarn, P., Yang, J.C., Chiu, Y.P.: The network effect on information dissemination on social network sites. Comput. Hum. Behav. **37**, 1–8 (2014)

8. Sun, J., Tang, J.: A Survey of models and algorithms for social influence analysis. In: Aggarwal C. (eds.) Social Network Data Analytics, pp. 177–214. Springer, Boston (2011). https://doi.org/10.1007/978-1-4419-8462-3_7

9. Gao, Q., Sun, C., Yang, C.: The influence of network structural properties on information dissemination power in microblogging systems. Int. J. Hum.-Comput. Interact. **30**(5), 394–407 (2014)

10. Ghosh, R., Lerman, K.: Predicting influential users in online social networks. Computer Science (2010)

11. Newman, M.E.: A measure of betweenness centrality based on random walks. Soc. Netw. **27**(1), 39–54 (2005)

12. Valente, T.W.: Network interventions. Science **337**(6090), 49–53 (2012)

13. Bonacich, P.: Some unique properties of eigenvector centrality. Soc. Netw. **29**(4), 555–564 (2007)

14. Katz, L.: A new status index derived from sociometric analysis. Psychometrika **18**(1), 39–43 (1953)

15. Page, L., Brin, S., Motwani, R., Winogrod, T.: The pagerank citation ranking: Bringing order to the web. Stanford Digital Librarics Working Paper (1998)

16. Medley, A., Kennedy, C., O'Reilly, K., et al.: Effectiveness of peer education interventions for HIV prevention in developing countries: a systematic review and meta-analysis. Aids Educ. Prev. Official Publ. Int. Soc. Aids Educ. **21**(3), 181–206 (2009)

17. Borgatti, S.P.: Identifying sets of key players in a social network. Computat. Math. Organ. Theor. **12**(1), 21–34 (2006)

18. Guo, Y., Chen, H.: Microblog user ranking based on PageRank and Hadoop. Int. Conf. Inf. Eng. **49**, 1083–1085 (2014)

19. Girvan, M., Newman, M.E.: Community structure in social and biological networks. Proc. Natl. Acad. Sci. **99**(12), 7821–7826 (2002)

20. Freeman, L.C.: A set of measures of centrality based on betweenness. Sociometry **40**, 35–41 (1977)

21. Watts, D.J.: Small worlds: the dynamics of networks between order and randomness. Phys. Today **31**(4), 74–75 (2002)

22. Seyed, M., Hosseini, B., Ildar, N., Qu, Q.: Opinion leader detection: a methodological review. Expert Syst. Appl. **115**, 200–222 (2019)

23. Ma, N., Liu, Y.: SuperedgeRank algorithm and its application in identifying opinion leader of online public opinion supernetwork. Expert Syst. Appl. **41**(4), 1357–1368 (2014)

24. Lü, L., Zhang, Y.C., Yeung, C.H., et al.: Leaders in social networks, the delicious case. PLoS ONE **6**, e21202 (2011)

25. Kleinberg, J.M.: Authoritative sources in a hyperlinked environment. J. ACM (JACM) **46**, 604–632 (1999)

26. Weng, J., Lim, E.P., Jiang, J., He, Q.: TwitterRank: finding topic-sensitive influential Twitterers. In: Proceedings of the Third ACM International Conference on Web Search and Data Mining, pp. 261–270 (2010)

27. Saez-Trumper, D., Comarela, G., Almeida, V., Baeza-Yates, R., Benevenuto, F.: Finding trendsetters in information networks. In: Proceedings of the 18th ACM SIGKDD International Conference on Knowledge Discovery and Data Mining, pp. 1014–1022 (2012)

28. Richardson, M., Agrawal, R., Domingos, P.: Trust management for the semantic web. In: Fensel, D., Sycara, K., Mylopoulos, J. (eds.) ISWC 2003. LNCS, vol. 2870, pp. 351–368. Springer, Heidelberg (2003). https://doi.org/10.1007/978-3-540-39718-2_23

29. Tai, A., Ching, W., Cheung, W.: On computing prestige in a network with negative relations. Int. J. Appl. Math. Sci. **2**, 56–64 (2005)

30. Guha, R., Kumar, R., Raghavan, P., Tomkins, A.: Propagation of trust and distrust. In: Proceedings of the 13th International Conference on World Wide Web, pp. 403–412 (2004)
31. Chen, Y., Wang, X., Tang, B., Xu, R., Yuan, B., Xiang, X., Bu, J.: Identifying opinion leaders from online comments. In: Huang, H., Liu, T., Zhang, H.-P., Tang, J. (eds.) SMP 2014. CCIS, vol. 489, pp. 231–239. Springer, Heidelberg (2014). https://doi.org/10.1007/978-3-662-45558-6_21
32. Zhao, C., Wang, S., Li, D.: Exploiting social and local contexts propagation for inducing Chinese microblog-specific sentiment lexicons. Comput. Speech Lang. **55**, 57–81 (2019)
33. Huang, Y., Liu, Y., Li, Q.: Public policy simulation based on online social network: case study of chinese circuit breaker mechanism. In: Chen, J., Nakamori, Y., Yue, W., Tang, X. (eds.) KSS 2016. CCIS, vol. 660, pp. 130–139. Springer, Singapore (2016). https://doi.org/10.1007/978-981-10-2857-1_11
34. Valente, T.W.: Social Networks and Health: Models Methods and Applications, vol. 40, no. 2, pp. 235–236. Oxford University Press, Oxford (2010)
35. Degenne, A., Forsé, M.: Introducing Social Networks. SAGE Publications, California (1999)
36. Hanneman, R.A.: Introduction to social network methods. Department of Sociology University of California Riverside (2005)
37. Wasserman, S., Faust, K.: Social Network Analysis: Methods and Applications. Cambridge University Press, New York (1994)
38. Cachia, R., Centre, J.: Social computing: study on the use and impact of online social networking. Agron. J. **75**(4), 53–54 (2008)
39. Memon, N., Larsen, H.L.: Practical algorithms for destabilizing terrorist networks. In: Intelligence and Security Informatics, IEEE International Conference on Intelligence and Security Informatics Proceedings, pp. 23–24. IEEE (2006)
40. Saidi, F., Trabelsi, Z., Salah, K., Ghezala, H.B.: Approaches to analyze cyber terrorist communities: Survey and challenges. Comput. Secur. **66**, 66–80 (2017)
41. Zhang, B., Zhang, L., Mu, C., Zhao, Q., Song, Q., Hong, X.: A most influential node group discovery method for influence maximization in social networks: a trust-based perspective. Data Knowl. Eng. **121**, 71–87 (2019)
42. Borrego, C., Borrell, J., Robles, S.: Hey, influencer! message delivery to social central nodes in social opportunistic networks. Comput. Commun. **137**, 81–91 (2019)
43. Zhu, H., Yin, X., Ma, J., Hua, W.: Identifying the main paths of information diffusion in online social networks. Phys. A **452**, 320–328 (2016)
44. Salas, E., Grossman, R., Hughes, A.M., et al.: Measuring team cohesion: observations from the science. Hum. Factors: J. Hum. Factors and Ergon. Soc. **57**(3), 365–374 (2015)
45. Greer, L.L.: Group cohesion: then and now. Small Group Res. **43**(6), 655–661 (2012)
46. Casey-Campbell, M., Martens, M.L.: Sticking it all together: a critical assessment of the group cohesion–performance literature. Int. J. Manage. Rev. **11**(2), 223–246 (2009)
47. Martí, J., Bolíbar, M., Lozares, C.: Network cohesion and social support. Soc. Netw. **48**, 192–201 (2017)
48. Burt, R.S.: Closure as social capital. In: Lin, N., Cook, K., Burt, R.S. (eds.) Social Capital: Theory and Research, pp. 31–56. Transaction Publishers, New Brunswick (2001)
49. Soboroff, S.D., Drew, S., et al.: Group size and the trust, cohesion, and commitment of group members. PhD (Doctor of Philosophy) thesis, University of Iowa (2012)
50. Carron, A.V., Spink, K.S.: The group size-cohesion relationship in minimal groups. Small Group Res. **26**(1), 86–105 (1995)
51. Mullen, B., Copper, C.: The relation between group cohesiveness and performance: An integration. Psychol. Bull. **115**(2), 210–227 (1994)

Understanding Shifts of Public Opinions on Emergencies Through Social Media

Zhihua Yan[1,2] and Xijin Tang[1,2(✉)]

[1] Academy of Mathematics and Systems Science,
Chinese Academy of Sciences, Beijing, China
zhyan@amss.ac.cn, xjtang@iss.ac.cn
[2] University of Chinese Academy of Sciences, Beijing, China

Abstract. Social media have brought tremendous changes to the aggregation and propagation of the public opinions about emergencies. Public opinions spread more widely, and produce great influence on government, business, and daily lives. Hence, it is an important task to gain comprehensive understanding of the evolution of public opinions. Sina Weibo, one of the most popular social media in China, plays critical roles in the development of public opinions. This paper applies burst analysis toward emergencies with example helmet of incident using Sina Weibo data, and different stages are divided to reflect different foci of the public. Based on topic model, a visualizing approach is then proposed to illustrate opinions evolution, including the birth, death, splitting and merging of topics along the whole procedure of the incidents. Such kind of study aims to provide additional perspectives about the emerging and evolving public opinions in the social media.

Keywords: Topic modeling · Kleinberg's burst model · Public opinions · Social media · Sina Weibo

1 Introduction

The development of social media provides open environments for the public to create and spread news, and share attitudes about societal events. Nowadays most of emergencies are promoted by social media, such as Sina Weibo and Wechat, and produce greater influence on government, business, and daily lives. Understanding evolution of the public opinions of emergencies through social media is important to government, business, and scholars. The public opinions are composed of numerous topics. It is very difficult to accurately catch how and why topics evolve over time. First, social media contain huge volume of unstructured data, which are hard to be dealt with. Second, topics are always composed of complex evolution patterns, such as birth, death, splitting, merging and etc. Perceiving relationships between topics is a challenging work.

Research on topic evolution originates from topic detection and tracking (TDT), which aims to search and organize event-based topics from textual news media materials [1]. Topics are regarded as a set of news stories relating to real-world events. With the advent of social media, TDT is also widely used for research of evolution of

© Springer Nature Singapore Pte Ltd. 2019
J. Chen et al. (Eds.): KSS 2019, CCIS 1103, pp. 175–185, 2019.
https://doi.org/10.1007/978-981-15-1209-4_13

public opinions. Furthermore, topic detection and topic evolution are in the foci of text mining. Traditional scientometric methods, such as co-word networks and co-citation analysis, have been used for research of topic identification and topic evolution [2]. Many multi-disciplinary findings have been achieved to a variety of domains, such as the government policies [3], discipline development [4], pop music trends [5] etc. Compared with co-word networks and co-citation analysis, the hierarchical Dirichlet process, a generative probabilistic topic model, performs better [6].

Topic models have been applied in various fields, such as scientific literature analysis and public opinion research, to reveal the evolution of topics across large collections of documents. Topic evolution has been successfully applied to explore and predict the scientific research trends. LDA was applied to the ACL Anthology to discover the research topic threads in the field of computational linguistics from 1978 to 2006 [7]. Based on the topic model, fifty key topics from transportation research articles were covered, and some general research trends are identified [8]. The political agenda of the European Parliament was explored to unveil the plenary agenda and detected latent themes using a dynamic topic modeling approach [9].

Along with the wide use of social media, user-generated content (UGC) is becoming an important data source for topic evolution research. Lau et al. tracked emerging events and trending topics on Twitter using LDA [10]. Barua et al. used LDA to identify the main topics presented at Stack Overflow, which was a question and answer website about computer technologies, and the variation in topics over time [11]. Cao and Tang employed dynamic topic model to explore the temporal patterns of changing topics on Tianya Zatan Board of Tianya Club [12]. Jia and Tang expanded the original storyline generating method using a novel multi-view graph to gain the evolution of the public attention of risk event on Tianya Club [13]. Xu and Tang proposed an improved algorithm to identify the evolution of societal risk events using Baidu hot news search words [14].

The remainder of this paper is organized as follows. Section 2 describes the development of topic modeling, burst detection, topic popularity and similarity. In Sect. 3, the process of data collection and data preprocessing is introduced. Section 4 presents our findings on the shifts of public opinions of emergencies on Sina Weibo through burst analysis and Sankey diagram. Finally, Conclusions and future work are given in Sect. 5.

2   Related Works

2.1   Burst Detection in Topic Model

Professor Kleinberg and his colleagues proposed a model to detect burst pattern from news streams or social media streams [15, 16]. This model draws an analogy with queueing models, and suppose the news streams as discrete batches of documents. Events are composed of discrete batches, some are relevant, while some are irrelevant. A bursty event comes into being if the arrive of relevant documents changes.

Suppose $B = (B_1, \ldots, B_m)$ represents document with m batches, which contains r_t relevant documents and d_t documents in total. Let $D = \sum_{t=1}^{m} d_t$ and $R = \sum_{t=1}^{m} d_t$. In a two state Kleinberg's model, burst state is denoted by q_1 and non-burst state is denoted by q_0. The expected fraction of relevant documents of state q_0 and q_1 are set by $p_0 = R/D$ and $p_1 = sp_0$, where s is a scaling parameter. $\tau(i_t, i_t)$ is defined as the cost of transition from none-burst state to burst state. The cost of state sequence $\boldsymbol{q} = (q_{i_1}, \ldots, q_{i_m})$ is defined as follows:

$$\sigma(i, r_t, d_t) = -\ln[\binom{d_t}{r_t} p_i^{r_t} (1 - p_i)^{d_t - r_t}] \tag{1}$$

Kleinberg's model has been widely used to uncover burst events through news streams and Twitter streams [17]. This paper applies this model to detect burst states of topics generated from Weibo. We set the scaling parameter as 1.8, and the cost parameter as 1.

2.2 Topic Popularity

Modern statistical topic models originate from information retrieval, and provide a powerful tool to reveal latent semantic structures from large collections of text documents. Introduced by Blei et al., LDA is a three-level hierarchical Bayesian model based on bag of words model [18]. The documents are represented by topics, which are multinomial probability distributions over words.

In LDA, the ith topic's popularity can be calculated by averaging posterior probabilities [19]. Let $\theta_{d,i,t}$ be the posterior distribution of topic i at epoch t for document d, and M_t be the number of documents at epoch t, the topic popularity is defined as:

$$\overline{\theta_{i,t}} = \frac{1}{M_t} \sum_d \theta_{d,i,t} \tag{2}$$

By calculating the topic popularity over time, we obtain the evolution rules of topics and group the topics under different labels.

2.3 Topic Similarity

In topic models, topics are represented by probability distributions over words. Many classical similarity measures, such as cosine similarity, Kullback-Leibler divergence, Jensen-Shannon distance, Euclidean distance, have been proposed and widely used in natural language processing. Here, Jaccard similarity, which is defined as the size of the intersection divided by the size of the union of the sets, is chosen to compute the similarities. In practical application of LDA, we always select the most representative terms to represent the topic to reduce the noise information. Let V_t' denote the set of top words generated by LDA at epoch t, the definition of Jaccard similarity is as follows:

$$J\left(V_t', V_{t+1}'\right) = \frac{|V_t' \cap V_{t+1}'|}{|V_t' \cup V_{t+1}'|} \tag{3}$$

If V_t' and V_{t+1}' are both empty, $J\left(V_t', V_{t+1}'\right) = 1$. The range of $J\left(V_t', V_{t+1}'\right)$ is from 0 to 1, and a higher value means a stronger relationship between topics.

3   Data Collection and Preprocessing

3.1   Data Collection

In April 11th, 2019, a construction worker released a video on Sina Weibo to show serious quality problems of helmet. Amazingly, this video became very hot, and spread widely on the Internet. This incident is a typical social emergency triggered by social media nowadays. Sina Weibo has 4,620 million active users, and is one of the most popular social media in China. Enormous social events are exposed on Sina Weibo, and become social hot spots. Helmet incident only lasted about one month, and became a hot issue in April 22th, ten days after exposing, and more than sixty thousand relevant posts were posted or forwarded, as shown in Fig. 1. Nevertheless, just three days later, only 12,569 related Weibo posts were released as public attention shifted quickly. The heated discussions on Sina Weibo caused stress to governments and then inspection to the construction sites and helmet companies have quickly been taken in different provinces.

In order to gain insights to the evolution of public opinions in helmet incident, 137,549 Weibo posts were collected from April 17th, 2019 to May 5th, 2019. Most of Weibo posts were created between April 22th and April 24th.

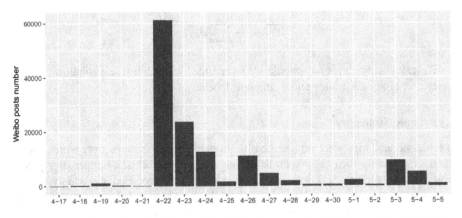

Fig. 1. Posts about helmet incident on Sina Weibo between April 17th and May 5th, 2019.

3.2   Data Preprocessing

Different from traditional news media, contents on Sina Weibo contains a large percentage of oral vocabulary and emoticons, which are difficult to be dealt with. As a consequence, we implement a comprehensive data processing on Sina Weibo dataset. Firstly, we remove user name and emoticons in Weibo content by regular expression.

Then, a Chinese-language stop word list containing "我们", "是", "所以", "然后", et al., and a reserved word list based on Baidu hot words [20] are employed to improve quality of word segmentation. These words which occur less than 5 are removed in helmet dataset [21]. As shown in Table 1, the refined corpus has 495,643 words, and the size of dictionary is 4034.

In order to model the evolving dynamics of topics, the helmet corpus is divided into five consecutive epochs. Each epoch contains four days except the interval from April 21th to April 23th for the sake of corpus balance.

Table 1. The statistics of helmet dataset after preprocessing.

Period	Microblog posts #	Corpus #	Dictionary #
4/17–4/20	1,421	5,221	692
4/21–4/23	84,949	341,033	3,929
4/24–4/27	30,025	99,676	3,390
4/28–5/1	5,189	15,020	1,781
5/2–5/5	15,965	34,693	986
Total	137549	495,643	4,034

4   Results and Analysis

4.1   Topic Discovering

In this study, the topic number parameter k estimated by perplexity is set to be 5. The experiment was done on a PC with 8 GB of RAM, and took approximately twenty minutes to run the LDA program[1]. To gain a comprehensive view of helmet incident, the whole corpus is trained by LDA, and five topics are generated. These topics covers all aspects of helmet incident, such as safety inspection, the changes of the worker's life, and so on. We label the topics manually, and use the top ten words to represent the corresponding topic, as shown in Table 2.

4.2   Burst Analysis

In emergencies, the foci of the public always shift with the development of incidents. Although we can acquire intuitive cognition through the variations of topic popularity, it is difficult to obtain accurate descriptions of the development of events. Hence, Kleinberg's model is employed to detect burst state in each topic and find salient topic in each time interval. Each topic has two states: burst state and non-burst state. The topic is in burst state when the arrival of Weibo posts changes greatly. Since the topics are always in non-burst state, we assume that topic receives much attention when the state of the topic changes to burst state. As shown in Fig. 2, the topic "safety inspection" was the most popular topic in the early stage of helmet incident, while the

---

[1] https://cran.r-project.org/web/packages/lda/.

Table 2. Topics generated by LDA.

No.	Label	Top-10 words
1	Social regulation	regulation (规训), issue (问题), cost (代价), position (位置), intrigue (密谋), consciousness (自觉), conspiracy (共谋), world (世界), means (办法), benefit (利益)
2	Safety inspection	helmet (安全帽), worker (工人), construction site (工地), safety (安全), quality (质量), leader (领导), company (单位), sudden inspection (突击检查), production (生产)
3	Cost of exposure	forwarding (转发), Sina Weibo (微博), comment (评论), Lu Xun (鲁迅), disappearance (消失), the mass (众人), skeleton (尸骸), society (社会), good people (好人), reality (现实)
4	Life of the worker	helmet (安全帽), video (视频), master worker (师傅), job (工作), disqualification (不合格), inspection (检查), unemployment (失业), the worker (当事人), livelihood (生活), quality (质量)
5	Product quality problem	issue (问题), resolution (解决), exposure (揭露), journalist (记者), truth (真话), society (社会), good end (好下场), incident (事件), brave (勇敢), gutter oil (地沟油), Sanlu milk powder(三鹿)

epoch from April 17th to April 21th was the burst period detected using Kleinberg's model. At this epoch, Weibo users paid more attention to the safety inspection of helmet manufacturers and construction sites, especially the efforts of inspection.

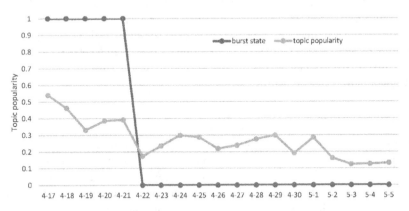

Fig. 2. Topic popularity and burst state sequence of the topic "safety inspection"

According to the analyses above, state sequences of topics generated by LDA are drawn on the timeline, as shown in Fig. 3. Along with the development of helmet incident, each topic moves from non-burst state to burst state at some epoch, and finally arrives at a non-burst state. For example, Topic 1 "social regulation" was in burst state from April 21th to April 22th, while Topic 4 "life of the person" was in burst state from April 18th to April 20th. Topic 3 "cost of exposure" was the hottest topic after April

24th. The transformation of topic state reflects the evolution of perception and focus of the public along this incident. The changes of public opinions about helmet incident are pretty obvious. In the earlier of helmet incident, the public focused on the helmet video, helmet quality, safety in construction site, life changes of the worker, who released the video to Sina Weibo. Then, the public began to rethink the cause of the event, and refer to similar product quality incidents, such as gutter oil incident, Sanlu milk power incident. In the later stage, the high cost of exposure in current society and the reasons why people were unwilling to expose social issues were discussed.

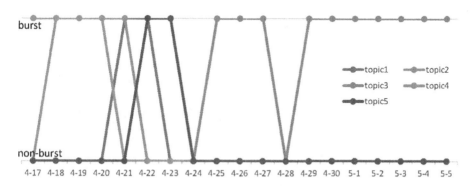

Fig. 3. State sequence of topics generated from helmet corpus in Weibo

4.3 Topic Evolving Dynamics

In the evolution of emergencies, topics not only emerge, develop and decline, but also split and merge [22]. To gain insights public opinions changes in emergencies, we propose a graphical approach based on LDA. Given the Weibo corpus D and the duration of emergency T, D is divided into chronological sequence $\{D_1, \ldots, D_n\}$ with the corresponding epoch sequence $\{T_1, \ldots, T_n\}$. Topics generated from corpus D are defined as global topics, while topics generated from corpus D_i are local topics. The algorithm of evolution graph is generated through the following steps:

Step 1: generate global topics and local topics using LDA. First, we use LDA to generate global topic set Z' based on corpus D and local topic set Z_i based on corpus D_i at epoch T_i.

Step 2: determine merging and splitting of topics. For topic $z_{i,j} \in Z_i$ and topic $z_{i+1,l} \in Z_{i+1}$, the similarity of $z_{i,j}$ and $z_{i+1,l}$ is computed by Jaccard similarity $J(z_{i,j}, z_{i+1,l})$. In order to reduce the impact of noise data, the most representative terms are selected to represent the topic. A threshold τ is used to determine whether there is a relationship between topic $z_{i,j}$ and $z_{i+1,l}$. Specifically, if $J(z_{i,j}, z_{i+1,l}) > \tau$, topics $z_{i,j}$ and $z_{i+1,l}$ are supposed have strong tie in term of topic content; if $J(z_{i,j}, z_{i+1,l}) < \tau$, topics $z_{i,j}$ and $z_{i+1,l}$ are supposed have no relation according to topic semantics.

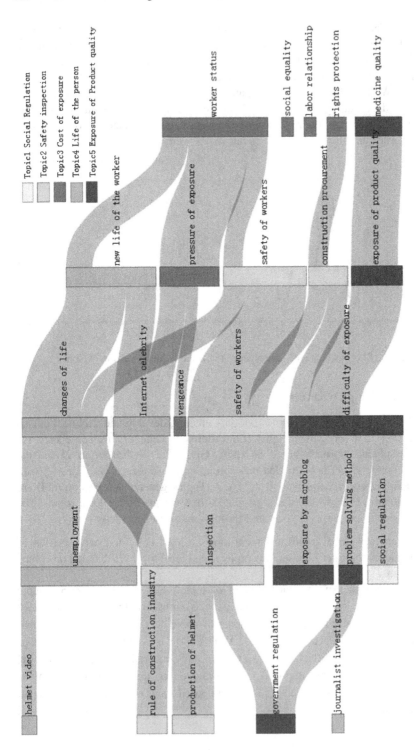

Fig. 4. An overview of public opinions shifts of helmet incident using Sankey diagram. The nodes represent topics at each epoch, and the width of edge represents the Jaccard similarity.

Step 3: determine topic lines. For topic $z_{i,j} \in Z_i$ and topic $z'_m \in Z'$, their similarity is computed by Jaccard similarity $J\left(z_{i,j}, z'_m\right)$. The threshold σ is employed to determine whether local topic $z_{i,j}$ is similar to global topic z'_m. If $J\left(z_{i,j}, z'_m\right) > \sigma$, topic $z_{i,j}$ is supposed to be similar to topic z'_m. Otherwise, topic $z_{i,j}$ is supposed to not be similar to topic z'_m.

Step 4: visualize the topic evolving process. We apply the Sankey diagram[2], a specific type of flow diagram to reveal transfers or flows within a complex system, to illustrate the topics evolution [23]. Topics are represented by nodes, while the relationships of topics are represented by the width of lines.

Here, we divide the helmet corpus into five epochs, as described in Sect. 3.2. The topic number parameter k is 5, because the sub-corpus contains fewer latent topics than the whole helmet corpus. Top 50 words in each LDA are selected as the representative terms, the threshold τ is set to 0.08 and σ is set to 0.11.

The Sankey diagram about the evolution dynamics of helmet incident is as shown in Fig. 4, where the x-axis represents time epochs, and the nodes represent topics at each epoch, and the width of edge represents the similarity between connected nodes. The nodes with the same color means that they are similar to the same global topic. For example, sub-topics "helmet video", "lose job", "life changes", "Internet celebrity" and "adapt to new life" are all about the changes of the worker's lives, and is similar to Topic 4 "life of the person". The Sankey diagram provides a visual representation of birth, death, splitting, and merging of sub-topics, which reveals the changes of netizen's foci. Sub-topics "rule of construction industry", "production of helmet" and "government regulation" merge into sub-topic "spot check", which is split into sub-topics "life changes" and "worker safety". Moreover, the Sankey diagram reflects the changes of foci of the public: Topic 1 "social regulation" receives attention only at the second epoch; Topic 3 "cost of exposure" are discussed from the third epoch, and become the hottest topic in the latter stage of helmet incidents; Topic 2 "safety inspection", Topic 4 "life of the worker" and Topic 5 "exposure of product quality" are discussed in the whole course of the incident.

5 Conclusions

This paper discusses the evolution of public opinions on emergencies using posts from Sina Weibo. Due to the huge volume of Weibo data, it is a challenge to build comprehensive views in order to explicitly describe the changes of public opinions. We employ topic model, burst detection algorithm and a novel Sankey diagram to provide a complete portrait of the social emergencies over social media. A real event, helmet incident, going highlighted via social media, is help to illustrate the whole analysis. The major contributions are summarized as follows:

---

[2] https://en.wikipedia.org/wiki/Sankey_diagram.

(1) Kleinberg's model and LDA are combined to detect the burst patterns of topics. Through the analysis of burst and non-burst states of topics, the emergencies are then divided into different development stages, to help gain insights of the shifts of the public's foci on emergencies.

(2) Sankey diagram based on LDA is proposed to uncover the topic evolution of emergencies. We use Jaccard similarity to represent the relationship of sub-topics and global topics, and generate a Sankey diagram which gives a precise description of emerging, decline, splitting and merging of the topics. Topic lines, which can uncover the content and temporal evolution of public opinions, are also described in Sankey diagram.

The Kleinberg's model and Sankey diagram generated from Sina Weibo data show a more distinct and comprehensive picture of shifts of online public concerns on emergencies. It is of great help to map out how the public opinions evolve clearly. However, we only illustrate the visualized analysis of the topics evolution. No comparisons are made with other visualized approaches toward topic evolution. Moreover, our approach is still under validation with large volume of streams of social media. More effective methods and automated analysis will be implemented in the future.

Acknowledgement. This research is supported by National Key Research and Development Program of China (2016YFB1000902) and National Natural Science Foundation of China (61473284 & 71731002).

References

1. Allan, J., Carbonell, J., Doddington, G., et al.: Topic detection and tracking pilot study: final report. In: Proceedings of DARPA Broadcast News Transcription and Understanding Workshop, pp. 194–218 (1998)
2. Leydesdorff, L., Nerghes, A.: Co-word maps and topic modeling: A comparison using small and medium-sized corpora (N < 1,000). J. Assoc. Inf. Sci. Technol. **68**(4), 1024–1035 (2017)
3. Rule, A., Cointet, J.P., Bearman, P.S.: Lexical shifts, substantive changes, and continuity in State of the Union discourse, 1790–2014. Proc. Natl. Acad. Sci. **112**(35), 10837–10844 (2015)
4. Lu, L.Y., Liu, J.S.: A novel approach to identify the major research themes and development trajectory: the case of patenting research. Technol. Forecast. Soc. Change **103**, 71–82 (2016)
5. Mauch, M., MacCallum, R.M., Levy, M., Leroi, A.M.: The evolution of popular music: USA 1960–2010. R. Soc. Open Sci. **2**(5), 150081 (2015)
6. Ding, W., Chen, C.: Dynamic topic detection and tracking: a comparison of HDP, C-word, and cocitation methods. J. Assoc. Inf. Sci. Technol. **65**(10), 2084–2097 (2014)
7. Hall, D., Jurafsky, D., Manning, C.D.: Studying the history of ideas using topic models. In: Proceedings of the Conference on Empirical Methods in Natural Language Processing, pp. 363–371. Association for Computational Linguistics (2008)
8. Sun, L., Yin, Y.: Discovering themes and trends in transportation research using topic modeling. Transp. Res. Part C Emerg. Technol. **77**, 49–66 (2017)
9. Greene, D., Cross, J.P.: Exploring the political agenda of the European parliament using a dynamic topic modeling approach. Polit. Anal. **25**(1), 77–94 (2017)

10. Lau, J.H., Collier, N., Baldwin, T.: On-line trend analysis with topic models: # Twitter trends detection topic model online. In: Proceedings of COLING 2012, pp. 1519–1534 (2012)
11. Barua, A., Thomas, S.W., Hassan, A.E.: What are developers talking about? An analysis of topics and trends in stack overflow. Empir. Softw. Eng. **19**(3), 619–654 (2014)
12. Cao, L.N., Tang, X.J.: Topics and trends of the on-line public concerns based on Tianya forum. J. Syst. Sci. Syst. Eng. **23**(2), 212–230 (2014)
13. Jia, Y.G., Tang, X.J.: Generating storyline with societal risk from Tianya Club. J. Syst. Sci. Inf. **5**(6), 524–536 (2017)
14. Xu, N., Tang, X.: Generating risk maps for evolution analysis of societal risk events. In: Chen, J., Yamada, Y., Ryoke, M., Tang, X. (eds.) KSS 2018. CCIS, vol. 949, pp. 115–128. Springer, Singapore (2018). https://doi.org/10.1007/978-981-13-3149-7_9
15. Kleinberg, J.: Bursty and hierarchical structure in streams. Data Min. Knowl. Disc. **7**(4), 373–397 (2003)
16. Leskovec, J., Backstrom, L., Kleinberg, J.: Meme-tracking and the dynamics of the news cycle. In: Proceedings of the 15th ACM SIGKDD International Conference on Knowledge Discovery and Data Mining, pp. 497–506, ACM (2009)
17. Takahashi, Y., et al.: Applying a burst model to detect bursty topics in a topic model. In: Isahara, H., Kanzaki, K. (eds.) JapTAL 2012. LNCS (LNAI), vol. 7614, pp. 239–249. Springer, Heidelberg (2012). https://doi.org/10.1007/978-3-642-33983-7_24
18. Blei, D.M., Ng, A.Y., Jordan, M.I.: Latent Dirichlet allocation. J. Mach. Learn. Res. **3**(5), 993–1022 (2003)
19. Griffiths, T.L., Steyvers, M.: Finding scientific topics. Proc. Natl. Acad. Sci. **101**, 5228–5235 (2004)
20. Hu, Y., Tang, X.: Using support vector machine for classification of Baidu hot word. In: Wang, M. (ed.) KSEM 2013. LNCS (LNAI), vol. 8041, pp. 580–590. Springer, Heidelberg (2013). https://doi.org/10.1007/978-3-642-39787-5_49
21. Blei, D.M., Lafferty, J.D.: A correlated topic model of science. Ann. Appl. Stat. **1**(1), 17–35 (2007)
22. Cui, W., Liu, S., Tan, L., et al.: TextFlow: towards better understanding of evolving topics in text. IEEE Trans. Visual Comput. Graphics **17**(12), 2412–2421 (2011)
23. Xie, W., Zhu, F., Jiang, J., et al.: TopicSketch: real-time bursty topic detection from Twitter. IEEE Trans. Knowl. Data Eng. **28**(8), 2216–2229 (2016)

Time to Liquidation of SMEs: The Predictability of Survival Models

Ba-Hung Nguyen[1,2(✉)], Galina Andreeva[2], and Nam Huynh[1]

[1] Knowledge Science, Japan Advanced Institute of Science and Technology,
Nomi, Japan
{hungba,huynh}@jaist.ac.jp
[2] Business School, The University of Edinburgh, Edinburgh, UK
{hung.nguyen,galina.andreeva}@ac.ed.uk

Abstract. One of the most crucial information in predicting firm time to liquidation is, given a firm has survived a certain period, its probability of surviving an arbitrary extended period. To provide updated results on the determinants of firm liquidations, we examine the survival of a large number of SMEs in the UK and follow them from 2004 up to 2016. We then compare the baseline hazard model and its updated hazard models on predicting time-to-liquidation using time-dependent performances, and for the latter one, we experiment with special of stratified bootstrap validation. We analyze the risk of going into liquidation of the UK SMEs from 2004 onwards using a baseline survival model and compare its predictive performance with the discrete survival model. A sample of 67,262 UK SMEs is employed in survival analysis with company fixed, demographic characteristics and time-varying financial elements. The results first show the significant effects of firm's demographic characteristics including number of trading addresses, number of directors, number of contacts, and number of subsidiaries. We also further stress on improvement in model accuracy using updated hazard models which utilized on the time-varying nature of firm's financial variables.

Keywords: SMEs · Survival model · Liquidation · Credit risk · Time-dependent performance

1 Introduction

In developing classification models for predicting corporate default, insolvency or liquidation, practitioners often rely on classifiers built on cross-sectional data and update it parameters periodically, e.g., 1 year, 3 years, or 5 years. This approach with advanced classifiers could achieve great results. However, one drawback of this approach is we discard a lot of censored information such as the firms that are not defaulted or liquidated at the end of the follow-up period. And since we have more behavioral data arising from firm business activities and their financial performance, which are usually updated quarterly or annually, there

© Springer Nature Singapore Pte Ltd. 2019
J. Chen et al. (Eds.): KSS 2019, CCIS 1103, pp. 186–200, 2019.
https://doi.org/10.1007/978-981-15-1209-4_14

is a shift toward employing the models that could be estimated using panel or longitudinal data with censored information. In this regard, little work has been done in the investigation the performance of survival models in term of predictive performance especially in credit risk modelling.

According to the National Federation of Self Employed & Small Businesses Limited (FSB), 2018, Small and Medium Enterprises (SMEs) accounted for 99%, 60%, and 52% of all private sector businesses, employment, and turnover, respectively. However, as the 2000–2017 trend of sustained growth was broken by a fall of 27,000 SMEs during 2017–2018, it raises a crucial question on the role of firms' characteristics, macroeconomics conditions, and their interactions. Moreover, the longitudinal study on insights of the determinants of such an increase in SMEs insolvencies plays a crucial role to provide a better risk assessment of the financial default during this severe economic and financial cycle.

In addition, few studies on this topic have been done, especially which covers the entire UK market while the Brexit onset is approaching. While the works of [12], and more recently, [13] shows limitation in the number of UK SMEs of the sample and in the financial covariates of these samples itself, Byrne et al., 2016 only focus on the pre- and onset- of the crisis period. To our knowledge, little study has examined the role of financial characteristics as time-varying covariates in the context of SMEs survival or in the aftermath of a financial crisis. [17] studied the behaviour of the UK SMEs from 2007 to 2010 - through the "credit crunch". In that period, they found non-financial variables are significant in predicting default similar to [3], more importantly, in spite of the fact that there are a raising number of default SMEs, they showed the stability and accuracy of credit risk models during that period.

Another crucial aspect in examining the performance of credit risk model is whether we focus on predicting cumulative default/liquidation applicants over a fixed future time interval versus predicting normal applicants over a range of follow-up times and whether applicant information is static or updated over time. There are many well-known statistical methods for evaluating prediction models with dichotomous outcomes, however, little has been done for survival outcomes taking into consideration of time-dependent sensitivity and specificity to fully characterize longitudinal predictor or models [9]. The methods we examine are particularly important in that they allow for an appropriate handling of censored outcomes commonly encountered with time-to-event credit risk data. In this work we extend the works of [17] and [13] in several ways. In terms of assessing the accuracy of the default prediction model, we include longer observations and time-varying covariates. In addition, we are able to investigate the time effects using discrete-time survival analysis, which deals with the imprecise time of the event. The remaining paper is presented as follows: Sect. 2 describes the related works in corporate credit risk and the application of survival analysis, we then detail our models and predictive measurements in Sect. 3. The data and empirical results are discussed in Sect. 4 and 5, respectively. We conclude with further research implications in Sect. 6.

2   Related Works

2.1   Corporate Credit Risk

The pioneer work on the predicting default risk is from Altman [1,2] where the authors used a set of financial ratios to determine the failure risk of corporates. In general, corporate credit risk modelling could be separated in three directions. The first direction, extending on the works of Altman, uses macro-economics index and a set of idiosyncratic factors. The second group, structural models of firm capital structure [15], regards a firm as default when its assets' value fall below the value of their liabilities. And the third direction is based on non-parametric, actuarial models widely used in the insurance industry, like Credit Suisse's CreditRisk+.

Research on the credit risk of SMEs using financial statements elements are not as popular as in large corporations since SMEs do not always have the full financial accounts and they are not necessarily required to report their financial statements. However, as stated above, SMEs account for more than 90% of the number of businesses in many markets, including the UK, hence, literature focuses on non-financial information like demographic features, or macroeconomics data. For a detail review on modelling failure of SMEs, reader could refer to [10] or [11].

2.2   Survival Analysis in Credit Risk

As SMEs often have shorter lifespan compare with other businesses, longitudinal study usually encounter censored data, in which SMEs drop out of the study because of several reasons, the directors retire or the firm become dormant, and so forth. Since cross-sectional study could not make use of censored data, survival analysis is becoming popular in SMEs longitudinal study, which is important in the SMEs insolvency research since we do not know the exact date of the event and/or the value of explanatory covariates that represent the SMEs characteristics at that event time.

Holmes [12] used a sample of 781 micro and SMEs firms in north-east England with the date of incorporation between 1973 and 2001 to assess the survival of new firms in the manufacturing industry. Recently, Kelly [13] examined the relationship of bank and non-bank credit and the macroeconomics in SMEs distress using long-run survival analysis, their study, which was carried on approximately 450,000 SMEs in Ireland, showed that macroeconomics, location, bank credit standards, and economic sector are determinants of firm survival. Byrne [6] used 9,457 UK, bank-dependent, and nonpublic firms with annual data from 2000 to 2009 to show that their survival chances are most determined by changes in uncertainty, especially during the 2008 global financial crisis. Macguinness [14] defined SMEs financial distress via z-score and examined if the trade credit help SMEs survive financial crisis using a panel data from 2002 to 2012 that included more than 8,000 UK SMEs.

Our study extends the work of [14] to cover longer time span that includes the financial crisis and examine the roles of time-dependent covariates, which originated from both financial and non-financial data, on the SMEs failure. More specifically, we assess the insolvency of the UK's SMEs using discrete-time survival analysis on a 12-year follow-up study. A sample of 67,262 SMEs with company fixed characteristics and time-varying financial variables were recorded from 2005 to 2016 time span. We also provide a comparison on capturing the behaviour of insolvent liquidations of the time-varying variables with firms' fixed characteristics.

3 Survival Models with Time-Dependent Covariates

3.1 Cox Proportional Hazard Model

Suppose we have a sample of n independent firms $(i = 1, .., n)$ and we start monitor each firm at a starting point $t = 0$ until time t_i. At this point, the firm i could be censored or an event occurs (liquidation or default). Censoring means that the firm is not monitored after time t_i because of lost to follow-up for some reasons (move to another market, stop business for retiring, and so forth). For each firm i we observe x_i, a $K \times 1$ vector of explanatory/independent variables or covariates.

To quantify how the occurrence of the event of interest depends on X, we usually use an unobservable/latent variable that specifies the occurrence/non occurrence of the event and we call it the **hazard rate**. Let a random variable T denotes the uncensored time of event occurrence and $\lambda(t)$ denotes the hazard rate, then:

$$\lambda(t) = \lim_{\Delta \to 0} Pr(t \le T < t + \Delta | T \ge t)/\Delta. \tag{1}$$

$\lambda(t)$ could be interpreted as the probability of an event occurs at time t given that it has not already occurred up to time t. If $\lambda(t) = \lambda$, it implies that T has an exponential distribution. Another way to represent $\lambda(t)$ is:

$$\lambda(t) = \frac{f(t)}{1 - F(t)}. \tag{2}$$

where $f(t)$ and $F(t)$ are the probability density and cumulative distribution function for T.

Cox [7] represents the hazard rate as a function of both time and covariates using a proportional hazards model,

$$\lambda(t, x) = exp(\alpha(t) + \beta'x). \tag{3}$$

where $\alpha(t)$ is a function of time and it could imply the distribution of T.

For the unobserved T of the censored firms, we could use the maximum likelihood (ML) for fully using of the information of these cases based on the likelihood equation:

$$L = \prod_{i=1}^{n} [f(t_i)]^{\delta_i} [1 - F(t_i)]^{1-\delta_i}. \tag{4}$$

where δ_i is a dummy variable with value equal to 1 if the firm is uncensored and 0 if censored. Equation (4) shows that an observation contribute a density function if time to event of this observation is observed and 1 minus cumulative distribution function if this observation is censored at time t_i.

Taking into consideration covariates vector x and from the Eq. (2):

$$L = \prod_{i=1}^{n} [\lambda(t_i, x_i)]^{\delta_i} [1 - F(t_i, x_i)]. \tag{5}$$

Also, from the Eq. (2):

$$F(t_i, x_i) = 1 - exp - \int_0^{t_i} \lambda(u, x_i) du. \tag{6}$$

And we can express the likelihood function entirely in the hazard rate:

$$L = \prod_{i=1}^{n} [\lambda(t_i, x_i)]^{\delta_i} exp - \int_0^{t_i} \lambda(u, x_i) du. \tag{7}$$

By this state, we could use ML estimation method to find the parameter of (7), however, this will lead to specifying the function α which is not a trivial task. Cox [7] proposed a Partial Likelihood (PL) method to estimate β without any restriction to $\alpha(t)$.

3.2   Discrete-Time Survival Analysis

Both ML and PL methods could be extended to incorporate multiple kind of events and the repeated events. The PH model could also be generalized to allow time-varying covariates and time-dependent coefficients ([18]). For incorporating the time-varying covariates, let x_{it} be the $K \times 1$ vector of explanatory variables that could take different values at different discrete times. The discrete-time hazard rate is defined as:

$$H_{it} = Pr[T_i = t | T_i \geq t, x_{it}]. \tag{8}$$

Using the logistic regression function to present hazard rate in term of time and explanatory variables, we have:

$$H_{it} = \frac{1}{1 + exp(-\alpha_t - \beta' x_{it})}. \tag{9}$$

or the logit form:

$$log \frac{H_{it}}{1 - H_{it}} = \alpha_t + \beta' x_{it}. \tag{10}$$

If there is an assumption that the data are generated by a continuous-time PH model like (3), the corresponding discrete-time hazard function is:

$$H_{it} = 1 - exp[-exp(\alpha_t + \beta' x_{it})]. \tag{11}$$

which has a solution as a complementary log-log (clog-log) function:

$$log[-log(1 - H_{it})] = \alpha_t + \beta' x_{it}. \tag{12}$$

(9) and (11) could be estimated by ML with the likelihood similar in (4):

$$L = \prod_{i=1}^{n} [Pr(T_i = t_i)_i^{\delta}][Pr(T_i > t_i)^{1-\delta_i}]. \tag{13}$$

Recall the conditional probability properties:

$$Pr(T_i = t) = H_{it} \prod_{j=1}^{t-1} (1 - H_{it}). \tag{14}$$

and

$$Pr(T_i > t) = \prod_{j=1}^{t} (1 - H_{it}). \tag{15}$$

Substituting (14) and (15) into (13) and taking the logarithm, we have the log-likelihood function:

$$log\, L = \sum_{i=1}^{n} \delta_i log \frac{H_{it}}{1 - H_{it}} + \sum_{i=1}^{n} \sum_{j=1}^{t_i} log(1 - H_{ij}). \tag{16}$$

We then can proceed to estimate $\alpha_t, t = 1..n$ and β. If we denote an event indicator y_{it}, $y_{it} = 1$ if sample i has an event at time t, then the (16) becomes:

$$log\, L = \sum_{i=1}^{n} y_{it} log \frac{H_{it}}{1 - H_{it}} + \sum_{i=1}^{n} \sum_{j=1}^{t_i} log(1 - H_{ij}). \tag{17}$$

which is basically a log-likelihood for the regression analysis of binary dependent variables.

3.3   Performance Measurements

Cross-Sectional Measurements: In classifying firms as having liquidation or not, a classifier is subject to 2 types of error: incorrectly classifying a liquidated firm as not having liquidation (false negative), leading to loss in investment, and, conversely, incorrectly classifying a normal firm as liquidation (false positive), subjecting the firm to unnecessary follow-up strictly scrutiny procedures. An efficient classifier should have high sensitivity (true-positive rate - TPR) and high specificity (1 minus false-positive rate - FPR). For a score scr and a threshold c, we define:

$$\text{Sensitivity}(c) = Pr(scr > c | \text{liquidated}). \tag{18}$$
$$\text{Specificity}(c) = Pr(scr \leq c | \text{normal}). \tag{19}$$
$$\text{AUC} = Pr(scr_i > scr_j | i : \text{liquidated}, j : \text{normal}). \tag{20}$$

the AUC represents the classifier's ability to rank liquidated firms above normal firms. An AUC of 0.5 indicates no discrimination whereas an AUC of 1.0 indicates perfect discrimination.

Time-Dependent Measurements: In the discrete-time settings, it is straightforward to extend the standard cross-sectional measurements to the survival context, where failure status is time dependent, that is to binarize the outcome at a specific time, t (30 days, 3 months, or 1 year), and define liquidations (L) as firms who liquidated before time t and normal (N) as those who remain event-free beyond t:

$$\text{Sensitivity}^L_{(c,\text{start}=s,\ \text{stop}=t)} = Pr(scr > c|T \geq s, T \leq t). \qquad (21)$$

$$\text{Specificity}^N_{(c,\text{start}=s,\ \text{stop}=t)} = Pr(scr \leq c|T \geq s, T > t). \qquad (22)$$

where T denotes survival time and s denotes the start time of failure firms. Then, the cumulative distribution of failure firms can be defined as firms who failure prior to t, as $T_i \in (s, t)$ and normal firm as those who are event free at time t, $T_i > t$.

Then the time-specific AUC, defined by the area under the time-dependent ROC curve for all value of FPR p, is as follows:

$$\text{AUC}^{L/N}(s,t) = \int ROC^{L,N}_{s,t}(p)dp. \qquad (23)$$

Here, AUC is the probability that a random firm i who has failed at time t has a larger score value than a random firm j who remains event free through time t, assuming both firms are event-free up to time t.

Finally, the concordance index or C-index and Sommer's D is defined as follows:

$$C\text{-index} = P(scr_j > scr_k|T_j < T_k). \qquad (24)$$

$$D_{xy} = 2C\text{-index} - 1. \qquad (25)$$

4   Data

4.1   Description

In this research we focus on the Small and Medium Enterprises (SMEs) in the UK, we could refer to [4] for the definition of SMEs in the UK and the liquidation state of the companies. According to Basel accord and European SMEs definition, SMEs are the enterprises that have the annual turnover less than 50 million of Euro, total assets less than 43 million of Euro, and number of employees less than 250, the number of subsidiaries is capped at 6, and the number of directors is 10 maximum. The data we get from Bureau van Dijk consist of 67,262 companies which have the date of incorporation between 01/01/2004 and 01/01/2005. The "Company status" recorded by FAME is presented on Table 1, where the majority of the enterprise are active:

Table 1. Company status

Process	Status	Count
0	Active (dormant), petition to wind-up	1
1	Active, meeting of creditors	6
2	Active, app. of liquidator	7
3	Inactive (no precision)	9
4	Active, petition to wind-up	12
5	Active, with vol. arrangement	27
6	Active, in administration	67
7	Active (dormant), in default	101
8	Active (receivership)	271
9	Active, in default	288
10	In liquidation	2484
11	**Active (dormant)**	10174
12	Dissolved	25481
13	**Active**	62207

Excluding the inactive company with no precision reasons, we regard a company as default if it is in the liquidation processes 0–10 according to UK government (https://www.gov.uk/liquidate-your-company). The dissolved category includes those that do not necessarily experience default or liquidation, they might stop operating because the owner retires, dies, or other reasons. We could consider this category as another level instead of only two level Liquidation/Active in further work. With the requirements of SMEs above, the number of SMEs from FAME that does not satisfy and is excluded is 5,617, among them, 299 SMEs are liquidated. Moreover, by setting an observation period of 12 years follow-up, we also exclude SMEs without exact date of incorporation, and we end up with the following number of default and active SMEs (Table 2):

Table 2. Company status

Code	Status	Count	Percentage
1	In Liquidation	1,598	2.38%
0	Active	65,661	97.62%

4.2 Imputation and Dummy Coding

As for the missing values of those numeric time-varying covariates, we impute missing values via multiple imputation using chain equations (MICE, [5]), which is a strongly recommended method by many research in longitudinal study [8].

There are several ways to compare the imputed data with the original one, including histogram, density plot, box plot, or quantile-quantile plot [16]. As we have longitudinal data, we present the scatter plots of imputed data and original data for continuous covariates. While the imputed values are not much different in Net Tangible Assets, Total Assets, and Current Assets compare with the original ones, in both mean and median values, the mean values of imputed data for Shareholder Funds are slightly lower than the complete data (Figs. 1 and 2).

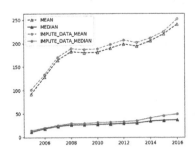

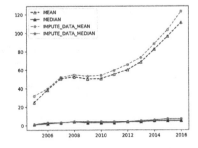

Fig. 1. Current Assets (th. GBP)   **Fig. 2.** Shareholders Funds (th GBP)

Figures 3 and 4 reveal the lower imputed values for Current Liabilities and higher values for Liquidity Ratios. They are reasonable since we expect the lower the Current Liabilities, the higher the Liquidity Ratios. This indicates the multiple imputation method using chained equations could capture the relationship between Current Assets and Current Liabilities with Liquidity Ratios.

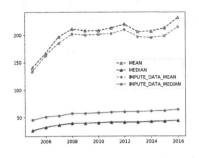

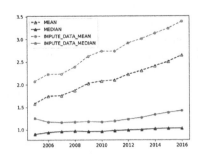

Fig. 3. Currents Liabilities (th. GBP)   **Fig. 4.** Liquidity Ratio (%)

The final longitudinal data consists of 800,076 SME-period observations with six fixed and six time-dependent covariates. Table 3 shows the binning of fixed, categorical covariates, and type of the time-dependent covariates as follows:

Table 3. All covariates

	Fixed covariate	Categories and Binning
1	#Directors	[≤ 3, > 3]
2	#Contacts	[≤ 2, > 2]
3	#Trading Addresses	[≤ 1, > 1]
4	#Subsidiaries	[0, > 0]
5	TAO[a]	Continuous
6	SIC[b]	[C, F, O][c]
	Time-Dependent covariate	Type
1	Total Assets (TA)	Continuous
2	Current Assets (CA)	Continuous
3	Net Tangible Assets (NT)	Continuous
4	Current Liabilities (CL)	Continuous
5	Shareholder Funds (SHF)	Continuous
6	Liquidity Ratio (LR)	Continuous

[a] The amount of money owed or due in a deposit account
[b] SIC: Standard Industry Code
[c] C: Construction; F: Food&Postal Activities; O: Others

After imputation, we examine the correlation and find Total Assets, Current Assets, and Current Liabilities are highly correlated, we remove TA in the final analysis. The distribution of the Liquidation SMEs over time adjusted for 12-year observation is shown in Fig. 5, which reveals, 2011, 2012, and 2015 have the highest number of liquidation SMEs.

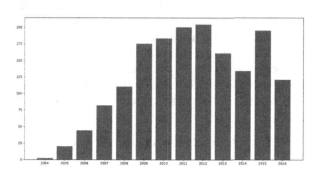

Fig. 5. Number of liquidations

5 Empirical Results

5.1 Baseline Hazard Model

We examine the performance of the baseline CoxPH model which is estimated based on the values of fixed covariates and time-varying covariates recorded in 2004 (Table 4):

Table 4. CoxPH

	Coef	Exp(Coef)	SE	z	p-value
Directors	0.4642	1.5907	0.0505	9.1855	0.0000
Contacts	−2.2186	0.1088	0.1692	−13.1143	0.0000
Trading_Addresses_2	0.3007	1.3508	0.0794	3.7893	0.0002
Trading_Addresses_1	−0.2894	0.7487	0.0587	−4.9337	0.0000
Subsidiaries	−1.4315	0.2389	0.2004	−7.1418	0.0000
TAO	0.1034	1.1089	0.0204	5.0766	0.0000
SIC_C	0.3925	1.4807	0.1312	2.9906	0.0028
SIC_O	−0.7723	0.4620	0.1177	−6.5599	0.0000
NT	0.0625	1.0645	0.0162	3.8486	0.0001
SHF	−0.0502	0.9511	0.0150	−3.3496	0.0008
LR	−0.3928	0.6752	0.0579	−6.7895	0.0000
CL	0.0226	1.0228	0.0138	1.6309	0.1029
CA	0.0603	1.0621	0.0143	4.2145	0.0000
LR	R^2	D_{xy}			
989.95	0.036	0.426			

LR is the Likelihood Ratio; R^2 is the Nagelkerke R-squared; D_{xy} is Somers' D.

Having more than two contacts, having zero or one trading address, and not belong to Construction or Food&Postal Activities are strongly significant and negatively affect the hazard ratio. Whereas belonging to the construction sector, possession of at least two trading addresses are significantly and positively affect the hazard ratio. The liquidation risk of firms which have more than two trading addresses or employ more than three directors or belong to the construction industry is higher compared with firms that have at most one trading address or have less than three directors or belong to Food&Postal Activities industry. In term of discriminatory power, the D_{xy} is 0.426 which translates to C-index of 0.713, an acceptable level. To further validate the results of this baseline survival model, we present the 10-fold bootstrap corrected values below:

Table 5. 10-fold bootstrapping validation

Index	Original sample	Training sample	Test sample	Corrected index	n
D_{xy}	0.4260	0.4289	0.4262	0.4233	10
R^2	0.0356	0.0361	0.0347	0.0341	10

The corrected values computed from 10 validation samples are all lower, yet not much different from the original sample. We then proceed to check the proportional hazard assumption using the Schoenfeld residuals test (Table 6):

Table 6. Proportional hazard test

	ρ	χ^2	p
Directors	−0.0039	0.0247	0.8752
Contacts	0.0310	1.5478	0.2135
Trading_Addresses_2	0.0170	0.4693	0.4933
Trading_Addresses_1	−0.0250	1.0063	0.3158
Subsidiaries	−0.0381	2.3925	0.1219
TAO	0.1388	3.2652	0.0708
SIC_C	0.0146	0.3428	0.5582
SIC_O	−0.0452	3.3112	0.0688
NT	−0.0051	0.0392	0.8430
SHF	−0.0364	1.5145	0.2185
LR	−0.0097	0.4190	0.5175
CL	0.0135	0.1307	0.7178
CA	−0.0427	1.3168	0.2512
GLOBAL		18.8643	0.1274

The proportional hazard assumption test above indicates the CPH's assumption is violated for all financial elements. As for firm demographic variables, this assumption is violated only for TAO and SIC_C variables at 10% significant level, this suggests further investigations on the interaction of these covariates with time or employing the their time-varying variants. However, at the current stage we only have the latest TAO as an only available value for each firm despite TAO is time-dependent. On the other hand, since firms rarely change their business sectors, it makes sense to keep SIC as fixed covariate.

5.2 Time-Varying Cox Proportional Hazard

We then fit the Cox PH model using the data with time-varying covariates. Table 7 presents the multivariate regression results and reports the same performance indicators as in Table 5. Since our data are extremely imbalanced, some validation samples might have singularity problem which causes the corresponding validation steps not converge, we mitigate this problem by preserving the ratio of liquidated firms/normal firms in each sample.

As for the signs of covariates, in firm fixed characteristics, all covariates remain significant. Having more than two contacts or one subsidiary significantly reduce the risk of liquidation compare with lesser contacts or no subsidiary at all. If a firm has more than three directors, its hazard almost 60% larger than that does not. Firms in the Construction industry have the higher hazard of liquidation. As for the time-varying covariates, the more Net Tangible Assets and Current Liabilities the lower the liquidation hazard. On the other hand, Shareholder Funds is not significant in this time-varying model.

Table 7. Time-varying CoxPH model

	Coef	Exp(Coef)	SE	z	p-value
Directors	0.4691	1.5985	0.0505	9.2798	0.0000
Contacts	−2.1668	0.1145	0.1684	−12.8642	0.0000
Trading_Addresses_2	0.3619	1.4361	0.0796	4.5483	0.0000
Trading_Addresses_1	−0.2579	0.7727	0.0591	−4.3672	0.0000
Subsidiaries	−1.3415	0.2615	0.2006	−6.6868	0.0000
TAO	0.1008	1.1061	0.0232	4.3375	0.0000
SIC_C	0.3112	1.3651	0.1319	2.3592	0.0183
SIC_O	−0.8955	0.4084	0.1176	−7.6138	0.0000
NT	−0.9177	0.3995	0.0485	−18.9020	0.0000
SHF	−0.0113	0.9887	0.0213	−0.5317	0.5949
LR	0.1336	1.1430	0.0123	10.8885	0.0000
CL	−0.8527	0.4263	0.0454	−18.7767	0.0000
CA	0.8834	2.4192	0.0451	19.6027	0.0000
LR	R^2	D_{xy}			
1009.26	0.032	0.466			

The discrete model's fitness is modestly lower compare with the baseline model. A high R^2 would require high precision in predicting the actual times of events, often an unrealistic goal. In fact, the values of R^2 in both models are suffered from long follow-up time and a limited number of events for the entire period (liquidated firms consists of roughly 2% of the total sample). Focusing on predictive power, we observe an uplift in the model accuracy in term of Sommers' D rank by incorporating the time-varying covariates into the discrete-time model. It gains approximately 2% in C-index (Table 8).

Table 8. Validation with sample stratification using bootstrapping

Index	Original sample	Training sample	Test sample	Corrected index	n
D_{xy}	0.4660	0.4819	0.4798	0.4640	10
R^2	0.0323	0.0330	0.0294	0.0306	10

By and large, despite SMEs do not have to disclose their detailed financial data, the survival analysis models reveal that SMEs' demographic covariates are significant. The performance of the baseline model is, though violates the proportional hazard assumption, acceptable by correctly ranking almost 3/4 pairs of cases. And by employing the time-varying variants of the financial covariates, we could not only relax the proportional hazard assumption for them in building the discrete-time survival model but also improve the model accuracy.

6 Conclusions

Our study extends the work of [12] to cover longer time span and focuses on the impact of using discrete-time approach in applying survival analysis for predicting business failure, especially on the combination of fixed and time-varying covariates, which originated from both financial and non-financial data of the SMEs. More specifically, using discrete-time survival analysis on a 12-year longitudinal study of 67,262 UK SMEs with company fixed, demographic characteristics, and time-varying financial variables, we first, despite the lack of financial data for the majority of SMEs, show the potential application in modelling SMEs liquidation using survival models, especially discrete-time survival models. Firm demographic features including the number of directors, number of subsidiaries, number of contacts, and number of trading addresses are all significant. And by including time-varying covariates, practitioners, on one hand, have a better understanding of how these covariates affect not only the event - the liquidation of SMEs, but also the time-to-event or time-to-liquidation. On the other hand, these time-varying covariates are significant and could improve the predictability of models.

There are apparent limitations in our study, first, we only focus on UK SMEs, which in turns, might behave very differently compared with other peers since the Brexit referendum in 2016. In addition, as SMEs data are prone to missing data, a comparative analysis on the effect of imputation methods on the performance of survival models for SMEs liquidation modelling is also expected. Another interesting avenue is to examine the competing risk models where we include the dissolved firms in the analysis. Finally, as our data follow SMEs from 2004 to 2016, other analysis on the pre-, post-, and onset of the 2008 crisis could reveal interesting results.

References

1. Altman, E.I.: Financial ratios, discriminant analysis and the prediction of corporate bankruptcy. J. Finance **23**(4), 589–609 (1968). https://doi.org/10.1111/j.1540-6261.1968.tb00843.x. http://doi.wiley.com/10.1111/j.1540-6261.1968.tb00843.x
2. Altman, E.I., Iwanicz-Drozdowska, M., Laitinen, E.K., Suvas, A.: Financial distress prediction in an international context: a review and empirical analysis of Altman's Z- score model. J. Int. Financ. Manag. Acc. **28**(2), 131–171 (2017). https://doi.org/10.1111/jifm.12053. http://doi.wiley.com/10.1111/jifm.12053
3. Altman, E.I., Sabato, G., Wilson, N.: The value of non-financial information in SME risk management. J. Credit Risk **6**(2), 95–127 (2010). https://www.risk.net/node/2160680
4. Andreeva, G., Calabrese, R., Osmetti, S.A.: A comparative analysis of the UK and Italian small businesses using Generalised Extreme Value models. Eur. J. Oper. Res. **249**(2), 506–516 (2016). https://doi.org/10.1016/j.ejor.2015.07.062. https://linkinghub.elsevier.com/retrieve/pii/S0377221715007183
5. Buuren, S.V., Groothuis-Oudshoorn, K.: mice: multivariate imputation by chained equations in R. J. Stat. Software **45**(3) (2011). https://doi.org/10.18637/jss.v045.i03, http://www.jstatsoft.org/v45/i03/

6. Byrne, J.P., Spaliara, M.E., Tsoukas, S.: Firm survival, uncertainty, and financial frictions: is there a financial uncertainty accelerator? Econ. Inquiry **54**(1), 375–390 (2016). https://doi.org/10.1111/ecin.12240. http://doi.wiley.com/10.1111/ecin.12240

7. Cox, D.R.: Regression models and life-tables. J. Roy. Stat. Soc. Ser. B (Methodological) **34**(2), 187–202 (1972). https://doi.org/10.1111/j.2517-6161.1972.tb00899.x. http://doi.wiley.com/10.1111/j.2517-6161.1972.tb00899.x

8. Dirick, L., Claeskens, G., Baesens, B.: Time to default in credit scoring using survival analysis: a benchmark study. J. Oper. Res. Soc. **68**(6), 652–665 (2017). https://doi.org/10.1057/s41274-016-0128-9. https://link.springer.com/article/10.1057/s41274-016-0128-9

9. Djeundje, V.B., Crook, J.: Dynamic survival models with varying coefficients for credit risks. Eur. J. Oper. Res. **275**(1), 319–333 (2019). https://doi.org/10.1016/j.ejor.2018.11.029. http://www.sciencedirect.com/science/article/pii/S0377221718309548

10. Gupta, J., Barzotto, M., Khorasgani, A.: Does size matter in predicting SMEs failure? Int. J. Finance Econ. **23**(4), 571–605 (2018). https://doi.org/10.1002/ijfe.1638. http://doi.wiley.com/10.1002/ijfe.1638

11. Habib, A., Costa, M.D., Huang, H.J., Bhuiyan, M.B.U., Sun, L.: Determinants and consequences of financial distress: review of the empirical literature. Acc. Finance (2018). https://doi.org/10.1111/acfi.12400, http://doi.wiley.com/10.1111/acfi.12400

12. Holmes, P., Hunt, A., Stone, I.: An analysis of new firm survival using a hazard function. Appl. Econ. **42**(2), 185–195 (2010). https://doi.org/10.1080/00036840701579234. http://www.tandfonline.com/doi/abs/10.1080/00036840701579234

13. Kelly, R., Brien, E.O., Stuart, R.: A long-run survival analysis of corporate liquidations in Ireland. Small Bus. Econ. **44**(3), 671–683 (2015). https://doi.org/10.1007/s11187-014-9605-1. http://link.springer.com/10.1007/s11187-014-9605-1

14. McGuinness, G., Hogan, T., Powell, R.: European trade credit use and SME survival. J. Corp. Finance **49**, 81–103 (2018). https://doi.org/10.1016/j.jcorpfin.2017.12.005. https://linkinghub.elsevier.com/retrieve/pii/S0929119917307484

15. Merton, R.C.: On the pricing of corporate debt: the risk structure of interest rates*. J. Finance **29**(2), 449–470 (1974). https://doi.org/10.1111/j.1540-6261.1974.tb03058.x. http://doi.wiley.com/10.1111/j.1540-6261.1974.tb03058.x

16. Nguyen, C.D., Carlin, J.B., Lee, K.J.: Model checking in multiple imputation: an overview and case study. Emerg. Themes Epidemiol. **14**(1) (2017). https://doi.org/10.1186/s12982-017-0062-6, http://ete-online.biomedcentral.com/articles/10.1186/s12982-017-0062-6

17. Orton, P., Ansell, J., Andreeva, G.: Exploring the performance of small- and medium-sized enterprises through the credit crunch. J. Oper. Res. Soc. **66**(4), 657–663 (2015). https://doi.org/10.1057/jors.2014.34

18. Tuma, N.B., Hannan, M.T., Groeneveld, L.P.: Dynamic analysis of event histories. Am. J. Sociol. **84**(4), 820–854 (1979). https://doi.org/10.1086/226863. https://www.journals.uchicago.edu/doi/10.1086/226863

Author Index

Printed in the United States
By Bookmasters